THE QUICK GUIDE TO

Wild Edible Plants

THE QUICK GUIDE TO
Wild Edible Plants

*Easy to Pick,
Easy to Prepare*

LYTTON JOHN MUSSELMAN *and* HAROLD J. WIGGINS

THE JOHNS HOPKINS UNIVERSITY PRESS

Baltimore

© 2013 Johns Hopkins University Press
All rights reserved. Published 2013
Printed in Canada on acid-free paper

Johns Hopkins Paperback edition, 2017
9 8 7 6 5 4 3 2 1

Johns Hopkins University Press
2715 North Charles Street
Baltimore, Maryland 21218-4363
www.press.jhu.edu

The Library of Congress has cataloged the hardcover edition of this book as follows:

Musselman, Lytton J.
  The quick guide to wild edible plants : easy to pick, easy to prepare / Lytton John Musselman and Harold J. Wiggins.
     pages cm
  Includes index.
  ISBN 978-1-4214-0871-2 (hdbk. : alk. paper) — ISBN 1-4214-0871-6 (hdbk. : alk. paper) — ISBN 978-1-4214-0872-9 (electronic) — ISBN 1-4214-0872-4 (electronic)
  1. Wild plants, Edible—East (U.S.) 2. Cooking (Wild foods)—East (U.S.) I. Wiggins, Harold J., 1953– II. Title.
  QK98.5.U6M87 2013
  641.3'03—dc23   2012027071

A catalog record for this book is available from the British Library.

ISBN-13: 978-1-4214-2429-3
ISBN-10: 1-4214-2429-0

Special discounts are available for bulk purchases of this book. For more information, please contact Special Sales at 410-516-6936 or specialsales@press.jhu.edu.

Johns Hopkins University Press uses environmentally friendly book materials, including recycled text paper that is composed of at least 30 percent post-consumer waste, whenever possible.

Book design by Kimberly Glyder

Contents

Acknowledgments vii

Introduction 1

    Wild Plants as Food 1

    Before You Begin 3

    Emergency Food 5

    How to Use This Book 6

    Guidelines for Using the Recipes 7

    About Flavorings, Sweeteners, and Oils 8

    Beverages 9

    Recipes for Failure 9

Deadly Harvest: Plants You Should Avoid 13

    Poison Ivy, Poison Oak, Poison Sumac 13

    Poison Hemlock 17

    Mushrooms 19

Nature's Storehouse of Edible Plants 21

Condiments 29

    Sassafras 29

    Field Garlic 31

Aperitifs 35

    Swamp Bay 35

    Red Spruce 37

Greens 41

    Chicory 41

    Curly Dock 45

    Glassworts 48

    Kudzu 50

    Stinging Nettle 53

    Black Walnut 56

Starches 63

    American Lotus 63

    Arrowhead 67

    Groundnut 69

    Nut Sedge 72

    Oaks 74

    Softstem Bulrush 78

    Spring Beauty 79

Grains and Plants Used Like Grains 83

    Cane 83

    Manna Grass 87

    River Oats 90

    Yellow Pond Lily 94

Flowers 97

    Black Locust 97

    Cattails 101

    Orange Day Lily 105

    Redbud 108

Sweets 111

    Indian Strawberry 111

    Pawpaw 114

Cordials 119

    Blueberries 119

Mushrooms 123

    Oyster Mushroom 123

    Chicken of the Woods 127

    Puffballs 128

Index of Recipes 133

Acknowledgments

With any project that extends over many years, there are numerous people to acknowledge for their help and support. We thank Bill and Denise Micks, Mike Hicks, Jennifer Alexander, Rick and Beth Blanton, Mary Hauffe, Sally Knight, and Jodie Wiggins. We are thankful for the support of our colleagues at Old Dominion University and the Army Corps of Engineers for their insight and helpful criticism. Among these are David Knepper, J. Paul Minkin, and Peter Schafran. Lloyd Hitchings aided with camera use.

Several generations of students have enthusiastically munched, gagged, enjoyed, and endured tasting various concoctions we have prepared. However, no undergraduates were harmed in these experiments.

Peter Schafran and David Cutherell critically read earlier drafts and improved the manuscript, and they suggested some recipes. However, any errors are ours alone.

We appreciate the cooperation of various landowners who allowed access to their property, as well as the staff and resources of the Eastern Shore National Wildlife Refuge; Great Dismal Swamp National Wildlife Refuge; Rappahannock National Wildlife Refuge; Upper Mississippi River National Wildlife Refuge; Au Sable Institute of Environmental Studies, Michigan; Cranberry Lake Biological Station, New York; First Landing State Park, Virginia; Kiptopeke State Park, Virginia; Merchants' Millpond State Park, North Carolina; Lake Phelps State Park, North Carolina; Pike Peaks State Park, Iowa; Wyalusing State Park, Wisconsin; Crow's Nest Natural Area Preserve, Virginia; and Loriella County Park, Virginia.

The research reported in this book would not have been possible without the support of the Hogan Endowment, which we gratefully acknowledge.

Our spouses and children have been a constant source of support and encouragement.

THE QUICK GUIDE TO
Wild Edible Plants

Introduction

ABOUT FIFTY BOOKS ON EDIBLE PLANTS covering most of North America are currently in print. Is another one needed? Many of these books emphasize the obvious food from plants, such as berries and nuts; some require considerable preparation and added ingredients when cooking. In some the recipes use main ingredients like sour cream or some other component that masks the flavor of the wild plants. Others draw on lore and purported uses, especially from Native Americans. Still others describe edible plants but give little indication of usable yield. While we include some of the best-known wild foods, as well as their indigenous uses, we believe our book is unique. Perhaps quirky best describes it.

This book deals with some of the most common plants of the Middle Atlantic states and the Northeast, many of which are underutilized. It is not a comprehensive treatment of edible plants in this region. Rather, this book tells the reader where, when, and how to locate and prepare plants—both native and introduced—and plant products overlooked or neglected by other sources. Both native and introduced wild plants are included. We have been guided by several general principles in the selection of plants.

First and foremost, the plants included are easy to identify and do not have any toxic look-alikes. It's also important to learn which plants cause dermatitis and poisoning. That is why we have included descriptions of poison ivy, poison oak, and poison sumac, as well as some of the most toxic plants, so they can be avoided. Second, plants to be collected must be common (not rare or threatened), so that collecting them will not be detrimental to their survival. Third, preparation must be simple, requiring readily available equipment and no more than three components per recipe. (Okay, we cheat on some recipes but only a few.) Last, we have drawn upon the use of plants in other cultures.

During our fieldwork we have nibbled, chewed, boiled, and baked any plant that is not documented as life threatening. We have also studied many plants reported in the literature to be edible but found them unpalatable, low yielding, or both. In many cases students have also eaten their way through the landscape with us. However, we don't recommend unbridled foraging for anyone without a working knowledge of the local flora, especially of rare and endangered plants.

Collecting and preparing food from wild plants is a labor of love, not necessity—the use of wild plants is more about fascination and discovery than it is about food. Major food plants are well known and extensively studied because of their importance to human cultures for millennia. There are many reasons our existence depends on these plants that are not native, and one is that they are adapted to agriculture and can produce acceptable yields. Wild plants, on the other hand, are not adapted to agriculture; put another way, they have not been subjected to artificial selection.

Consider a field of corn. Why are all the ears approximately the same size and at the same place on the corn stalk? The corn has been bred for uniformity of ripening and positioning of the ears for mechanical harvest. Compare this with a native grain, Cane. Should you be fortunate to find a stand of Cane fruiting, some grains will drop from the plant at a touch while others are still ripening. To gather a majority of the Cane grains, therefore, would require several visits, whereas a threshing machine will collect virtually all of the ripe corn in an hour or so because all the ears are at the same stage.

In the wild, plants are under natural selection, while crops are artificially selected for, meaning they are chosen for their agronomic traits rather than for traits allowing survival under non-cultivation. Left on their own, crops such as corn cannot compete, and a field left unharvested will yield only a few corn plants the next year, the majority of which will bear little resemblance to their proud, straight parents. Wild plants vary wildly; cultivated plants are tame and homogenous.

This variability manifests itself in several ways. One is bitterness. Many wild plants—including acorns, water lily rhizomes and buds, some wild grains, and a diversity of other plants—often contain high concentrations of very bitter substances. These are not harmful in small quantities, but in larger doses they do interfere with both palatability and digestibility. Boiling in several changes of water will often help remove these substances, which are usually tannin compounds. The plants can also be placed in a solution of water and wood ashes, which will remove a lot of the tannins. As their name implies, tannins have been used in tanning because they bind the proteins in hides and thereby preserve them.

Oxalates are another group of compounds common in plants. Their role in the life of the plant is uncertain, but they may be waste products of metabolism. Plants are much more discrete about waste management than are animals—plants have no excretory system comparable to those of animals. While the distribution of oxalates is widespread, there are some families, such as the Arum family (Araceae), with high concentrations of oxalates. One group from that family that has been documented to cause swelling of the mucous membranes so severe that it blocks breathing is Dumbcane (*Diefenbachia* species); sometimes this swelling has had fatal results. Native relatives of Dumbcane, such as Jack-in-the-Pulpit (*Arisaema triphyllum*), will also make the most loquacious speaker dumb if sufficient quantities are ingested.

Oxalates have adverse effects on the kidneys and are often the main component of kidney stones. Boiling the plant may destroy the oxalates. In fact, there are many valuable wild foods that contain oxalates. Don't eat them raw, however. Oxalates will usually produce a sharp burning sensation in your mouth, which can last for hours. Some will even irritate skin because the long, needle-like crystals, called *raphides,* puncture the skin.

In recent years considerable attention has been paid to the accumulation of nitrates in plants. This is of concern to the wild food enthusiast because of several well-known edibles, including the Pigweeds (*Amaranthus* species) and Goosefoot (*Chenopodium album*). We have eaten both but hesitate to recommend these because they could be collected from fields receiving large amounts of nitrogen fertilizer. True, to be harmed would require considerable intake of these tainted plants, but we prefer to err on the side of caution.

Plant chemistry varies and people vary as well. Some people are simply more susceptible to gastric upset from plants than are other people. Others of us have allergies. If you are sensitive, avoid unknown wild plants or, if you wish, taste only a very small quantity.

BEFORE YOU BEGIN

DON'T START DOWN THE TRAIL of harvesting plants from nature until you take a few precautions regarding collecting and processing. There is a misconception that plants collected in the wild are pure and healthy. Unfortunately, this is not always true. For this reason you should be acquainted with some necessary precautions in wild food foraging.

There is a good possibility that some of the plants you harvest, especially those that are weeds in agricultural crops, in urban areas, or along roads may have been sprayed with toxins. Usually there is no way for you to determine whether the plant has been sprayed. As a precaution, do not collect where it is obvious that spraying has occurred, as in power line rights-of-way, golf courses, and similar highly maintained areas. You should avoid an area if you see "burning" or discoloration on the leaves of nearby trees and shrubs. Second, thoroughly wash whatever you collect. Obviously, in more remote areas there is less likelihood that plants will have been sprayed with chemicals.

Some areas may have toxins in the soil. Before the ban on the use of leaded gasoline, roadside plants often had considerable lead in their tissues, which is just one example of how plants can take up toxins. Even though leaded gasoline was banned in the 1990s, trees and other long-lived plants many still retain lead in their tissues. Another example is Sacred Lotus (*Nelumbo nucifera*), a water lily that is a common vegetable in Southeast Asia and is widely planted in the United States. It accumulates cadmium, which even in trace amounts can be toxic. Plants collected on or adjacent to industrial waste sites may be unsuitable for harvest because they contain heavy metals and chlorinated hydrocarbons.

Not all poisons, of course, have human origins. Plants and fungi produce an extraordinary number of chemicals, including toxins. Some fungal toxins are extremely poisonous substances and are carefully screened for in commercial food

Salt Marsh Cord Grass (Spartina alterniflora): *The hard black structures are the "horns" of Ergot Fungus. This stage of the toxic fungus appears in late summer and autumn.*

production. Of these, aflatoxins produced by the fungus *Aspergillus flavus* are perhaps the most serious. To lessen the chances that these toxins are present, collect only fresh seeds and fruits, not those that have been on the ground for some time. For example, when collecting acorns, it is best— though not easiest—to pick them off the tree rather than gathering them from the ground. Hot, dry weather promotes this fungus, which produces a yellow-grey or green growth. Discard any material that is discolored.

A second fungal toxin is the ergot fungus, *Claviceps purpurea,* which is also very poisonous. It was the cause of Saint Anthony's fire of the Middle Ages, when people were poisoned by eating flour contaminated with ergot. It may also have been the source of presumed witchery in colonial Salem, Massachusetts. Derivatives of this fungus are found in such modern psychotropic drugs as methylergometrine and ergotamine.

While the common grain rye is the most frequent host among culti- vated grains, ergot does infect several native grasses. The overwintering body of the fungus forms in the developing ovary of the grass so that the fungus and the grain are mature at the same time. A hard, black struc- ture extending hornlike from the grain appears late in the summer or early fall. When the grain is threshed, the horn of the fungus is broken off and included with the grains. Fortunately, modern milling techniques remove the fungus. Be sure to examine very carefully any wild grasses you plan to eat so you can avoid this dangerous fungus.

Ultimately, the most important guideline is common sense. Know the plant you are dealing with and where it came from. Be certain of its identity. Learn which parts are edible and how they are prepared. Be very cautious about experimenting by yourself. If you are a serious wilderness camper, know which plants are suitable emergency foods. Be familiar with the litera- ture on edible wild plants. Safe preparation and practice is the key so your adventure in learning will produce safe, satisfying results.

EMERGENCY FOOD; OR, "I THINK I SHALL NEVER SEE A MEAL AS LOVELY AS A TREE"

WHAT SHOULD YOU DO IF you are lost and without food? The best thing to do is climb a tree and look for a fast food outlet because survival food is neither fast nor tasty. But should you find yourself in that situation, we recommend two foolproof sources of emergency food. Again, we stress *emergency.* If you are starving, precious calories need to be conserved.

If you are in a forest in eastern North America, you are almost certainly near one of the native firs, hemlocks, pines, or spruces. All of these have edible, though hardly haute cuisine, inner bark, where a layer of living cells is present all year-round. If you have an ax, slit the bark of the tree and peel it off. If needed, a rock can be used. Then, using the blade of the ax, scrape the sap both on the outside of the tree where you have removed the bark and on the inside of the piece of bark you have removed. Gooey? Overtones of turpentine? Yes, but this material includes the cells of the tree that produce new growth—in other words, embryonic tissue—which contains proteins and starches that may give you ample energy to find real food. The advantage of food from a tree is that it is a resource any time of the year, including winter. We have eaten the sap from Red Spruce (see photo, opposite) and, although you might not serve it at a dinner party, the taste bordered on pleasant.

Should you be near a marsh where the ubiquitous Cattails are likely to be found, you have a dependable source of good-quality starch—assuming, of course, that it is not the middle of winter in the northern states and the water and soil are frozen. Simply excavate the rhizomes just at or below the surface of the mud. They are filled with starch and can be eaten raw, although they are better boiled or roasted.

HOW TO USE THIS BOOK

MAKE SURE YOU HAVE PROPERLY identified the plants you are using for food. Eastern North America is blessed with numerous authoritative wildflower books. Our book has a limited selection of plants, none of which are rare or unusual and all of which are easy to identify. However, if you need a usable, reliable aid for identifying herbaceous plants, we can recommend *Newcomb's Wildflower Guide,* by Lawrence Newcomb and Gordon Morrison (Little-Brown).

We have imprecisely divided the book into sections based on the type of food produced so that grains and their kin are together, greens together, and so forth. To get a general overview read the introductory material before dealing with the specific plants. Within each section, plants are arranged by their common name. How common are the common names? There are no formal guidelines for the construction, use, and uniformity of common names, so we simply use the common names that are common to us. For precision and accuracy we include scientific names. They are essential and provide the means of appropriately seeking further information on the plant in question.

A description gives the diagnostic features of the plant necessary for proper plant identification. Then we describe the collection and preparation

Red Spruce: Students at Cranberry Lake Biological Station prepare the inner bark of Red Spruce for food. Scraping the outer portion of the trunk and the inner part of the bark yields a slimy but nutritious food.

of the edible part of the plant in more detail. Photographs taken especially for this book aid in identification and illustrate some of the foods prepared from the plants. Last are the recipes. Treatments of each plant vary considerably in length simply because some plants are more versatile in their utility than others.

GUIDELINES FOR USING THE RECIPES

OUR CRITERIA FOR SELECTING the plants to include are noted above. For preparation, stark simplicity is the rule of thumb that has guided us in crafting these recipes. This means that (1) the targeted plant must provide the majority of the food (except flavorings), not just be an adjunct; (2) preparation should be simple, without any specialized equipment; and (3) the recipe should have only three to five ingredients. We have personally foraged, prepared, and eaten all the plants and foods in this book.

Obviously, there are some reasonable parameters regarding just how natural the preparation should be. For example, a food dehydrator is easier than sun drying and a food processor is easier than pounding with a stone. And a coffee grinder or food processor is easier than a traditional quern or mortar and pestle for grinding food.

Concocting these recipes drew upon our passion for plants more than our culinary expertise. Although we have consulted the corpus of information on wild edibles, all of the recipes, except those for *mok* and Mesopotamian pollen paddies, are original.

Use your own imagination in crafting recipes with the plants in this book as well as others (with all precautions, of course!). Our starkly simple recipes can be expanded with the addition of other ingredients, including combinations of other plants we describe. This book is only the beginning of the journey, more of a compass than a road map. Think of it as an introduction that can lead to your own experimentation.

ABOUT FLAVORINGS, SWEETENERS, AND OILS

ONE OF THE PROBLEMS we have faced is how to season these wild plant foods. There are few native plants that produce anything like an easily prepared spice or condiment. We have included a few condiments as examples of what can be done with these limited resources, giving us a new appreciation of why Europeans were so anxious to have access to the spices from the East.

Two of the most basic items in any kitchen are vegetable cooking oil and sugar. Although both are derived from plants, we include little information on their preparation from native plants. To do so we would have to prepare the equivalent of an auto repair manual for plants with attendant specialized equipment. Sugar and oil, however, do deserve passing attention.

What about natural sweeteners? The first natural sweetener that comes to mind is, of course, honey, which itself is a plant product made with the help of bees. But sugars directly from plants require more effort. In eastern North America there are several trees that can produce delicious sweeteners, Sugar Maple (*Acer saccharum*) being the best known. But other maples as well as Birches (species of *Betula*) can be tapped for their sugar-rich sap. We have not included techniques for using tree sap for sweeteners because of the effort and specialized equipment necessary to collect even a small amount. There are several helpful websites with instructions for sugaring, including www.umext.maine.edu/onlinepubs/pdfpubs/7036.pdf.

Native Americans derived fats from animals, which is the easiest way to obtain the end product of plant compounds. However, buffalo fat does not loom large in our recipes. We have done some preliminary oil extraction from Black Walnuts, but this requires a complex process with special

equipment. The earnest food forager can pursue sugar and oil from nature while the rest of us go to our kitchen cabinets.

WE MENTION A FEW BEVERAGES. Virtually any edible leaf can be made into a tea, as a quick look in a trendy health food store reveals. A separate section on teas is not included. Since we include chicory for its greens, we have included the preparation of a delicious coffee substitute from this common weed. Because they are so simple to produce, we include two aperitifs and one cordial.

RECIPES FOR FAILURE

THE TWO MOST IMPORTANT ASPECTS of our book are our foraging for the plant in nature and our recipes. In other words, this book is predicated on original research. We knew there was scant literature on certain plants and how much edible product they would yield compared to better-known plants. So we set out to study these lesser-known plants.

By canoe or on foot we crossed icy streams in winter and tenacious miasmic mud in scorching summer heat, clambered through redoubts of thorns and spines, climbed trees, battled clouds of insects—often without a recipe. Other times we spent days extracting plants from muck that seemed to extend to Middle Earth, carved out the purported edible parts, boiled, parched, ground, baked, and dehydrated only to find that what remained was a taste that would gag a maggot. Many of our great expectations to forage for alleged edible plants became premeditated disappointments.

One example is our foraging for Southern Wild Rice (*Zizaniopsis milicea*), a member of the grass family that is reputed to be edible and is common in the southern half of the United States. We found this heavily armored plant in freshwater tidal marshes and gathered several bags of grains, rewarded with lacerations on our arms from the sawtooth leaf margins. The grains were painstakingly threshed and winnowed, but we did not get enough starch from the grains to make a recipe for an easily prepared meal. Despite references asserting that this plant is edible, there just wasn't much to eat.

Another example is our recipe attempts for Arrow Arum (*Peltandra virginica*), a wetland plant found at the edge of marshes, rivers, and lakes. There are numerous references by early colonists suggesting the use of

Arrow Arum by Native Americans. So we harvested large quantities. No matter how we tried to boil, wash, grind, soak in wood ashes, and dry the Arrow Arum rootstock, the results were always unsuccessful. Moreover, this was a painful lesson as the calcium oxalate crystals, or raphides, penetrated our skin. Even after days of preparation, the starchy slices of root caused a burning sensation on the lips and mouth and even on our hands.

Native Passion Flower (*Passiflora incarnata*), also known in the South as Maypops, have fruits with a pleasant fruity sour taste when ripe and were once eagerly collected by rural children as a snack. So we tried to make a cordial from them. For reasons inexplicable, the taste was insipid to mildly disgusting. We have little passion for that use of the fruit.

Of the three water lilies native to eastern North America, the White Water Lily (*Nymphaea odorata*) is perhaps the best known. It occurs all across the continent and is abundant, easy to identify, and therefore a natural candidate for this book. There are reports of preparation of starch from the massive rhizomes as well as food from the flower buds. The image of this lovely, fragrant ornament of fresh water, however, belies the awful taste of its parts. We harvested the rhizomes and soaked, baked, boiled, and roasted them. The incredible tannins remained. Wines are often described as having soft, round tannins. White Water Lily has flinty, lacerating tannins. Same for the buds. One method people in earlier generations used to remove the bitter tannins was soaking in wood ashes. So we duly burned several logs, collected the ashes, put them in water along with the buds, and left them for several days. No change in the horrid taste. Since native plants can vary widely in the concentration of chemical constituents, we repeated these trials with plants from North Carolina, Virginia, and New York. No change in the horrid taste.

Perhaps no native plant of eastern North America is more easily identified than the widespread and appropriately named Devil's Tongue Cactus (*Opuntia humifusa*). The attractive fruits portend a fruity taste, perhaps as a worthy drink, and the pads are a good source of nutrition. The extreme slime present in both the fruit and the pad, however, render Devil's Tongue unpalatable.

There are more failures—hips (fruits) from a native rose that have spine-covered seeds, greenbrier species with supposed edible rootstocks that yield the equivalent of sawdust in texture and taste, Pickerel Weed leaves so fibrous they are better beaten than eaten, lichens that might be edible but have sand that is impossible to remove—which we have experienced and will spare the reader the details. We believe that it is vital to mention failures in our book because it is an important way to learn about foraging and trying new recipes. Even the failures can lead to success in knowing which "edible" plants are a waste of time. Experimentation and patience is needed in finding new edible plants. It is our hope that our readers will learn from our experiences and find their own recipes from nature!

Deadly Harvest: Plants You Should Avoid

ANYONE WHO IS A WILD FOOD forager needs to recognize the most seriously harmful plants they might encounter. Throughout much of the United States, this means Poison Ivy (*Toxicodendron radicans*) and its lesser-known evil cousins, Poison Oak (*T. pubescens*) and Poison Sumac (*T. vernix*); all of these cause dermatitis. We also include two of the most deadly plants to ingest, Poison Hemlock (*Conium maculatum*) and Water Hemlock (*Cicuta maculata*).

POISON IVY, POISON OAK, POISON SUMAC

THE DERMATITIS-CAUSING SUMACS need to be carefully distinguished from the common red-fruited sumacs. Present-day botanical classification places the dermatitis-causing plants into the appropriately named genus *Toxicodendron* ("poison tree" in Latin) and the other sumacs, well known for their edible fruits, in the genus *Rhus*. As a result, older literature will, for example, refer to Poison Ivy as *Rhus radicans* and more recent writing will refer to *Toxicodendron radicans*.

All parts of Poison Ivy, Poison Oak, and Poison Sumac cause dermatitis in any season and should be avoided. The chemical compound that causes the blistering and swelling is urushiol. Other plants in the same family as Poison Ivy, the Anacardiaceae, include Mangos, Pistachios, and Cashews, all of which contain small amounts of urushiol in their unripe fruits and vegetative parts. This explains why some people have food allergies to these well-known products.

All three species of *Toxicodendron* have alternate compound leaves. The leaves of Poison Ivy and Poison Oak have three leaflets, and those of Poison Sumac have seven or nine leaflets. Poison Ivy is a woody vine with compound leaves up to 6 inches long. The stem of Poison Ivy is adapted for high climbing in trees and has specialized roots that attach to the bark.

Poison Ivy in midsummer: Each leaf consists of three leaflets. The terminal leaflet is symmetrical, but the lateral leaflets are not—a good diagnostic feature of this dangerous plant. Developing fruits are seen at the middle right.

Flowers are green and inconspicuous; fruits are white berry-like structures that ripen in late summer.

Where can you expect to find Poison Ivy? Just about anywhere you are looking for wild edibles. It has a remarkable breadth of habitats from ocean dunes to mountain forests, margins of marshes to urban settings.

Unlike its very similar relative, Poison Oak is not as widespread nor as frequent as Poison Ivy and is often confused with it. It prefers drier habitats and is a small shrub that never climbs high on trees. What distinguishes Poison Oak is its overall hairiness and the distinctive lobing of the leaflets, which can bear an amazing resemblance to oak leaves. It is, however, no relation to true oaks (species of *Quercus*). Nor is it related to the common English Ivy (*Hedera helix*). Poison Ivy leaflets can also be lobed but never to the extent of Poison Oak. The leaves are the same size and the flowers and fruits are superficially similar.

 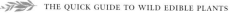

Poison Ivy in late winter: The numerous hairs (aerial roots) growing off the thick, woody stem are a good way to recognize Poison Ivy in the winter. The winter buds are rusty brown (lower center).

Last in the lineup of these villains is Poison Sumac. Fortunately, in most of its range through the eastern United States it is not very common nor does it form extensive stands—unlike its congeners—except in peat bogs in the upper Midwest. A shrub to 15 feet tall, it has the most specialized habitat preferring wetlands. And the leaves of Poison Sumac are much different than Poison Ivy and Poison Oak. Like its relatives the leaves are alternate in their arrangement on the stem but rather than three leaflets. Poison Sumac has 7–11, always an odd number with the leaflets (not the leaves) opposite one another on the leaf stalk. The leaf stalk is dark red, a consistent identification feature.

Be on the alert for these wicked plants. Fascinating in their own way, they should nevertheless be avoided, since even brushing against them or being exposed to smoke when they are burning can cause dermatitis in sensitive people. You may encounter Poison Ivy almost anywhere; Poison Oak is less widespread, and Poison Sumac is seen very infrequently.

Poison Oak: several of the leaflets resemble oak leaves, the source of the common name.

Poison Sumac with developing fruit shows large leaves with numerous leaflets and red leaf stalks.

POISON HEMLOCK (*Conium maculatum*) is probably the single most toxic plant in the United States. It is a native of Eurasia but is now found all over the United States and Canada. With its height and masses of white flowers, it is an attractive plant of wet roadsides, margins of rivers and lakes, irrigation ditches, and other moist areas, where it forms populations of hundreds and sometimes thousands of plants. That's a lot of poison in one place, considering that a tablespoon of the root is purported to be enough to kill an adult man.

This plant killed Socrates, who was forced to drink a decoction of Poison Hemlock. In its cocktail of toxins is coniine, a central nervous system poison. The affect of this poison is illustrated in Socrates' account of his own poisoning, when he related how his limbs became numb.

All parts of the plant are highly toxic but, because it somewhat resembles in odor its relatives carrot and parsnip, Poison Hemlock has been mistakenly ingested. Like many other members of the carrot family (the Apiaceae), Poison Hemlock has leaves that are finely divided, and the base of the leaf wraps around the stem. The stem is mottled with purple blotches, reflected in the name *maculatum,* which means spotted.

Poison Hemlock is apparently a biennial, like carrot, and dies after producing fruit, which resemble celery seeds or cumin, its well-known relatives. Large masses of dome-shaped flowers are produced in the spring and attract numerous insects. This is a vigorous plant that grows to 10 feet tall.

Water Hemlock (*Cicuta maculata*) is not as notorious as Poison Hemlock but is just as dangerous. It occurs in all 48 states and Alaska, as well as most provinces of Canada. Water Hemlock is more specific in its habitat requirements than is Poison Hemlock, and it is seldom found in large populations. Along with its better-known relative, this is one of the most toxic plants in North America, and it has been implicated in the poisoning of grazing animals. Like Poison Hemlock, it has a carrot- or parsnip-like smell that belies its danger.

These two plants are perhaps the most deadly of those you will encounter, but many others can cause serious illness and even death. Remember the old saying that any plant is edible—once. Never eat an unknown plant, and bear in mind that there are numerous plants with both edible and toxic parts.

Poison Hemlock: (a) the leaves resemble those of garden carrot as well as Queen Anne's Lace (Daucus carota). *(b) Purple blotches on the stem characterize this species.*

Water Hemlock in its typical habitat at the edge of a river: The leaf segments are narrower than those of Poison Hemlock.

"THERE ARE OLD MUSHROOM HUNTERS and there are bold mushroom hunters but there are no old, bold mushroom hunters." Wise counsel! But any wild food forager will naturally want to consider using mushrooms, so we include three that are widespread and easy to identify along with the caveat to never, ever even try a mushroom that you have not properly identified. And proper identity requires more than following folk guidelines such as growing in a circle, darkening a silver spoon, and other fallacies.

NEVER EAT AN UNKNOWN MUSHROOM! If you suspect mushroom poisoning, collect the mushroom, taking care to get the underground parts by digging, not pulling. Put it in a paper bag (some mushrooms will self-digest in a plastic bag) and have it examined by a specialist. Oftentimes the Poison Hotline at 800-222-1222 can direct you to someone who can identify the mushroom so that proper treatment can be sought.

Nature's Storehouse of Edible Plants

THERE ARE MANY EXAMPLES of edible plants that the novice can easily identify and collect by simply walking through forests and fields, paddling lakes, rivers, and streams, or perusing your backyard. Before going out into the field, water, or woods, learn as much as you can about plants you intend to find as food and their unique characteristics. If you have to eat something and you are on the move, survival foods can make a big difference when you are away from modern conveniences, even your camping area. It is good to know that these edible foods, while they may not be tasty, can provide enough energy to keep you going in situations such as being lost or other emergencies. Many plants are edible only during certain times of the year. The authors have struggled with the idea of preparing a treatment for a simple meal or snack from the smorgasbord of easily available edible plants that individually or collectively would make a downright great salad! This will take practice and a working knowledge of a plant's regional availability, identification, and abundance. Most of the plants we recommend in this section are found ready for collection in early to late summer.

Walking along the edge of a hardwood forest could yield a bounty of young Common Greenbrier (*Smilax rotundifolia*) tendrils. They grow in late spring to early summer and are crisp and appealing to start munching on as is. This is a great survival snack or, if enough can be collected, a stand-alone meal.

Usually occurring in open fields, you may also encounter Dandelion (*Taraxacum officinale*) leaves, along with discovering Field Garlic (*Allium vineale*) bulbs, a sprig of Yellow Wood Sorrel (*Oxalis europaea*) flowers, and the Common Blue Violet (*Viola papilionacea*). With little or no preparation you have a wonderful, nutritious salad with these early summer plants.

As you continue your foraging adventures, you may find many other plants on a tree, growing as a vine, or waiting to be pulled from the ground. The leaves of Curly Dock (*Rumex crispus*) can be eaten raw. Curly Dock leaves are found throughout spring, summer, and fall. Along

Field Garlic (Allium vineale) *bulbs.*

Blue Violet (Viola papilionacea) *flowers.*

Yellow Wood Sorrel (Oxalis europaea) *flowers.*

  THE QUICK GUIDE TO WILD EDIBLE PLANTS

Chickweed (Stellaria media).

Redbud (Cercis canadensis) *flowers.*

American Lotus (Nelumbo lutea) *seeds.*

roadsides and in fields and waste places you will find the tender stems and leaves of Chickweed (*Stellaria media*), which can also be eaten raw along with the leaves of Yellow Sweet Clover (*Trifolium officinalis*). In the same habitat, do try a few sprigs of Purslane (*Portulaca oleracea*) for good measure.

As you advance into the hardwood forest, throw in the fruits of Patridgeberry (*Mitchella ripens*). Along the trail or at the margin of the forest, don't forget to add something sweet to complement your salad, such as the ripe berries of the many species of blackberries and raspberries (*Rubus* spp.) and Mulberry (*Morus alba*). Wow! We are now the hunter-gatherers of the new age!

Many edible flowers are found along roadsides in the spring and summer. For example, those of Day Lily (*Hemerocallis fulva*), Redbud (*Cercis canadensis*), and Black Locust are welcome additions to any meal.

Black Locust flowers are slightly crunchy and sweet, with a pleasant aroma. You will see the pendulous Black Locust flowers hanging in great masses from the tree in spring through midsummer. Closer to the edge of rivers and streams look for the flowers of Elderberry (*Sambucus canadensis*) and Swamp Rose (*Rosa palustris*). These flowers are just delicious. Both of these plants yield edible flowers in midsummer.

During your late fall exploration you can pick the fruits of Persimmon (*Diospyros virginiana*) from the low-hanging boughs of trees along the banks. Persimmons are best eaten immediately after the first frost.

In still water in lakes, freshwater estuaries, and marshes you can find the seeds of American Lotus (*Nelumbo lutea*) in mid- to late summer. These seeds can be easily plucked from the flower heads, peeled, and eaten right on the spot. Near the margin of the lake or wetlands are Cattails (*Typha latifolia*), which can be pulled from the mud, peeled, cut with a knife, and eaten any time of year.

Condiments

FEW OF OUR NATIVE PLANTS provide suitable condiments. A condiment is something used to enhance the flavor of food. We usually think of spices and herbs in this context. A spice is usually derived from a fruit or seed, and an herb comes from leaves. While not precise, this is a helpful working definition. The few condiments we discuss would be, in a more technical sense, herbs.

Traditionally, native peoples used more than just plant-derived materials to flavor their food. Salt is one obvious example. The desire to enhance quotidian meals was the basis of the spice trade, the pursuit of which altered world history. Our examples are less cosmic in their lore but do improve the palatability of basic wild plant foods.

SASSAFRAS

SASSAFRAS (*Sassafras albidum*) has a special place in American lore and is best known for making Sassafras tea, produced by harvesting the bark of the roots just before leaf break, in earliest spring. The venerable drink has fallen into disrepute because of the link between ingestion of large amounts of Sassafras tea and possible carcinogens.

Found throughout eastern North America, Sassafras is usually a shrub or small tree 10–12 feet tall. Under ideal conditions it can grow to 30 feet. The bark is, appropriately, a root beer brown. But it is the leaves that are most distinctive because of the variability present on each plant—some leaves are two-lobed, some three-lobed, and some unlobed. The characteristic aroma, like root beer, is found in all parts of the plant. The fragrant essential oils of Sassafras are like other members of the laurel family (Lauraceae) with such redolent members as Bay, Camphor, and Cinnamon.

Sassafras (Sassafras albidum): *on this specimen, most of the leaves have three lobes.*

Filé (PRONOUNCED FEE-LAY)

Collect Sassafras leaves just as they emerge in early spring. There is only a small window of opportunity for collection as the larger leaves lack the desired flavor. Pick these early flushes of leaves, taking care not to include the harder scales at the base. The scales are hard to chew. Dry in the sun or with a dehydrator. Store in a closed container away from heat and light.

FIELD GARLIC (*Allium vineale*) is a Eurasian introduction to North America, where it is now abundant over a wide area, especially in areas with mild winters. Like its relative commercial Garlic (*Allium sativum*), it is a perennial and in the onion family (Alliaceae). The Field Garlic bulb is covered with a thick brown skin. It tastes and smells almost exactly like cultivated garlic, except much stronger. Field Garlic can be substituted in practically any recipe requiring cultivated garlic.

Onions and garlic are reported to have health benefits, such as lowering cholesterol and blood pressure, as well as anti-bacterial and anti-inflammatory properties. Like most members of its family, Field Garlic produces allyl sulfide compounds that give onions and garlic their characteristic sharp smell. The active incredient, allicin, is formed when the plant is crushed. In Field Garlic the allicin may be more concentrated and can cause some intestinal upset if too much is eaten raw. For this reason, use smaller amounts in recipes than you would with garlic.

When developing Sassafras leaves are about 1½ inches long, it's time to collect them to make filé.

Field Garlic is relatively easy to find, with its tall, distinctive tubular leaves up to 3 feet tall. To collect, simply grasp the tall stems of this plant at the base and pull up the bulbs. Shake off excess dirt. Usually, where one plant is found there are likely to be dozens nearby.

Field Garlic (Allium vineale): *the leaves of Field Garlic appear only in the winter and early spring; then the plant dies back to the bulbs, making it impossible to locate.*

Field Garlic Powder

20 Field Garlic bulbs

Carefully cut away the outer brown, fibrous layers from each bulb. Then cut each bulb into small pieces and put them into a food dessicator to dry for a day or more. It is essential to keep the garlic dry because the preparation will easily absorb water and then be difficult to grind. Alternatively, place the small pieces of Field Garlic bulbs on a baking sheet and dry in an oven at 200°F until hard and dry. Then grind in a mortar and pestle. This may take some time, as the dried pieces are usually hard and difficult to grind. The effort is worth it, since you now have one of nature's tastiest condiments.

Aperitifs

BAY LEAF, THE COMMON HERB so frequently used in Mediterranean dishes, is a member of the laurel family (Lauraceae), which also includes such aromatic plants as Cinnamon, Sassafras, Camphor, and Swamp Bay. Swamp Bay (*Persea palustris*) and the closely related Red Bay (*Persea borbonia*) inhabit acid swamps in the southeastern United States. A simple way to distinguish between them is by the hairy undersurface of the leaf of Swamp Bay. Because of the similarity of the two species, we are treating them together.

Swamp Bay (Persea palustris) *in winter, showing its evergreen leaves: the leaves are more flavorful if they are collected later in the year.*

These bays are much-branched shrubs up to 12 feet tall with thick, leathery evergreen leaves that are fragrant when crushed. Leaves are glossy on the upper surface and up to 5 inches long and 1½ inches wide. Flowers are small and borne in the spring; in the late autumn they produce a small oval fruit one inch long, dark blue and fleshy. Although the resemblance is not evident, the common avocado (*Persea americana*) is in the same genus.

Swamp Bay provides one of the few sources of seasoning among our native flora. The flavor is not unlike that of the commercial bay leaf (*Laurus nobilis*).

Swamp Bay in early winter: the leaves are evergreen, with a shiny upper surface (a) and a densely hairy undersurface (b).

Pick the leaves late in the season. Bear in mind that the flavor intensity varies considerably from one shrub to the next. For this reason, collect from as many different shrubs as possible. Dry the leaves in the sun or with a dehydrator. After drying, the leaves can be stored in a sealed container and kept at room temperature. As with all spices, avoid storing near heat and light. The pungency of the leaves will decrease over time, so it is best to collect new leaves each year. For a recipe using Swamp Bay, see Arrowhead Faux Bay Leaf (page 69).

∿ RECIPE ∿

Swamp Bay Aperitif

Nothing could be simpler than preparing an aperitif from this aromatic plant. Just place four leaves in one cup of vodka and age for at least three months. Collecting the leaves late in the season will produce a more flavorful brew.

The flavor? Intense, with cumin overtones and a short finish.

Swamp Bay Aperitif

RED SPRUCE

A SIGNATURE TREE IN PARTS of the Adirondacks, adjacent Canada, and New England, Red Spruce (*Picea rubens*) is a conifer, valued for its timber. As with many other plants of the northern forest, the range of red spruce extends down to the southern Appalachian mountains. A Red Spruce tree can live for several hundred years. Its green leaves, or needles, are stiff and about an inch long. Cones, usually only found in the upper boughs of the tree, hang down and are about 2½ inches long.

Red Spruce (Picea rubens)*: A young Red Spruce tree showing the characteristic branching (a). The stiff, dark-green needles are borne on brown twigs (b).*

Two other wide-ranging spruce species can occur with Red Spruce and be confused with it. Black Spruce (*Picea mariana*) is generally restricted to bogs. The twigs can be distinguished by their brown hairs, and the needles are dark green. The third species, White Spruce (*Picea glauca*), is a truly boreal species extending all across the North American continent. In the northeastern United States it favors wet areas along streams but not bogs. The needles are very stiff and have a whitish layer of wax on the undersurface.

Black Spruce is sometimes used in the production of beer. A few twigs of Black Spruce may be thrown into the mash to add flavor. The odor of White Spruce (described as resembling cat urine) has not fostered much interest in its use for flavoring.

Collecting

The best time to collect Red Spruce twigs is in the late summer so that the leaves of the current season have had time to develop their characteristic flavor. Because the intensity and quality of the flavor can vary from tree to tree, select material from different trees, if possible.

RECIPE

Red Spruce Aperitif

1 cup vodka

Place the twigs directly into the final bottle. Pour in vodka and seal.

It will take several months for the flavor to develop. Our three-year-old preparation is very flavorful. This makes a great aperitif for the Christmas season. The taste is piney but delicate with a generous finish.

The flavor? Intense, with cumin overtones.

Red Spruce Aperitif: the aperitif after six months of aging. The flavor and color are best after three years.

Greens

GREENS ARE THE SIMPLEST of all wild foods to harvest and cook. Because they are leaves, they are a good source of vitamins and fiber but provide little in the way of carbohydrates. And they are seasonal, which limits their utility. However, greens can be dried and stored in closed containers for use during the winter. Older literature refers to greens as *potherbs,* an accurate description as they are usually prepared by boiling. We discuss several that are very widely distributed, abundant, and easy to prepare.

CHICORY

CHICORY (*Cichorium intybus*) is an erect, somewhat woody, perennial herbaceous plant up to 5 feet tall, with bright blue flowers. Varieties of Chicory are cultivated for salad leaves. Chicons, for example, are blanched buds. The roots are baked, ground, and used as a coffee substitute or coffee additive. Sky-blue flowers that open only in the morning make this an attractive weed. Its leaves bear a resemblance to Dandelion leaves and, like Dandelion, Chicory has a milky juice.

The use of Chicory extends back to the ancient Egyptians, who considered it an aphrodisiac because of the resemblance of its milky juice to semen. Chicory was probably introduced into western Europe as an addition to coffee. The same may be true of North America, where it is now a common naturalized weed in just about any disturbed area. It is an attractive weed, however, and a stand of a dozen plants in flower is striking.

Young Chicory shoots are sold as a vegetable in several Mediterranean countries. Chicory is also grown as a forage crop for livestock. However, perhaps the best-known use for Chicory in North America is as a coffee substitute. Its roots are roasted and pulverized into a brown powder. There is nothing like a cup of hot, fresh Chicory around a campfire at night!

Chicory (Cichorium intybus): *Chicory flower head.*

Collecting

Collecting Chicory roots takes some effort. You will need a pair of heavy gloves, a spade or shovel, and a sharp knife. Grasp the stem of the plant with gloved hands and pull the entire Chicory taproot out of the ground. This may require a spade or shovel. Use a sharp knife to cut away the roots from the stems. To make 4 cups of Chicory coffee, you will need 10–15 roots.

*Blanched Chicory leaves ready for harvest (a), freshly dug Chicory roots (b),
and roasted ground roots (c).*

Chicory Coffee Substitute

10–15 Chicory tap roots

Carefully wash the roots under running water to remove dirt and grit. A small scrub brush or even an old toothbrush will do the job. Alternatively, you can peel the roots, though this is not necessary. Dry the roots. Lay the Chicory roots on a cookie sheet lined with aluminum foil and bake at 350°F for 45 minutes until the roots are brown. If you are camping, roast the roots in a saucepan over an open fire. The important thing is to get the roots hard and brown so they can be easily pulverized. When the roots are roasted and cooled, put them in a coffee grinder and pulverize into a fine powder. In our experience, Chicory coffee is surprisingly close in taste and aroma to the real thing.

Winter Chicory Blanched Leaves

6 Chicory roots

In late summer dig the roots (per instructions above). Thorough washing, however, is not needed. Place the roots in a plastic container (approximately 3 x 4 feet) filled with garden sand, which is available at a garden shop. Some so-called "sand" is actually ground limestone—do not use this. Place the box of sand in a dark place. Within a week, white, succulent leaves will emerge and can be used in salads. Leaves will continue to emerge for about a week to replace those you cut. The taste will resemble true endive (*Cichorium endiva*) but may be more bitter. You can have a continual supply of fresh blanched Chicory for much of the winter by keeping a supply of the roots in the refrigerator to replace those you have harvested.

CURLY DOCK (*Rumex crispus*), also known as Yellow Dock and Narrow Dock, is in the buckwheat family (Polygonaceae) and is native to Eurasia. This family, also known as the knotweed family, contains some members that are edible, such as Buckwheat (*Fagopyrum esculentum*), Rhubarb (*Rheum rhaponticum*), and Sea Grape (*Coccoloba uvifera*).

The mature Curly Dock plant is easy to spot. It is usually 3 feet high and has small, dark brown, winged one-seeded fruits atop the tall erect stems. Curly Dock has small, stalked, green or rust-colored flowers (¼ inch long) along a long, slender branching cluster. The mature fruit are small brown seeds (called *achenes*) surrounded by three wings, or small flaps, with smooth margins.

Curly Dock has long, narrow leaves with distinctive curled or wavy margins; the leaves grow from a basal rosette. Paper-like membranous sheaths wrap themselves around the joints of this plant; this is characteristic of the buckwheat family. There are smaller leaves near the clusters of inconspicuous flowers at the top half of the plant. We have used the roots to prepare a yellow dye. Flowers appear June through September, often going to seed as early as June.

Both the leaves and seeds of Curly Dock are edible. The leaves can be blanched, steamed, boiled, or chopped raw and added to salad, although the leaves may taste bitter when raw. The seeds of Curly Dock can be ground into flour, which can be used as a substitute for whole-wheat flour. The difference is that Curly Dock flour is gluten free and is darker brown. To prepare Curly Dock flour and meal, strip the seeds from the stem and grind the seeds and chaff into a flour.

Curly dock (Rumex crispus) *has tall, erect, brown stems. In midsummer these dark brown plants contrast with the surrounding green vegetation.*

Curly dock (Rumex crispus)*: note the brown achenes surrounded by three wings, or small flaps, with smooth margins.*

The young, tender leaves of Curly Dock are best in winter and early spring.

Curly Dock seeds can be separated from the chaff with your fingers.

Curly Dock / Buckwheat Pancakes

1 cup Curly Dock flour

3 tsp. olive oil

1 egg

2 tsp. sugar

¼ cup water

Prepare a skillet with the olive oil and heat on high. Beat 1 egg with the sugar in a medium bowl. Add 1 cup Curly Dock flour and beat well. Pour in ¼ cup water. Mix all together until well blended. With a large spoon ladle small pancakes into the skillet. Turn each until cooked dark brown. This yields enough Curly Dock / Buckwheat pancakes for 6 servings.

Curly Dock Cookies

1 cup Curly Dock flour

1 egg

1 tsp. baking soda

1 Tbsp. sugar

Mix all ingredients together. Hand mold cookie patties and place onto a cookie sheet lined with aluminum foil. Bake in oven preheated to 300°F for 30 minutes.

With a little work and patience, Curly Dock seeds can easily be processed into nutritious flour. First, locate a stand of several mature Curly Dock plants along the margins of a field or road. The seeds and chaff may be stripped away from the stem with your fingers.

Put the seeds and chaff in a 1-gallon bag. Crush the bag several times with your hands to release as many seeds as possible from the chaff. Take the crushed seeds and chaff out of the bag, and pour or push them through a large tea strainer to gather more seeds. Winnowing in a breeze will also help separate grain and chaff. Remove leaf parts and small stems. Put the remaining seeds into a coffee grinder and grind to make flour. It is all right to have some chaff in the mixture, especially if you prefer more fiber in your diet.

GLASSWORTS

GLASSWORTS (*Salicornia* species) occur along the coasts of North America, inland salt marshes, and alkali lakes. They have a unique appearance, with highly modified leaves, small flowers, and the remarkable ability to live in high salinity. This affinity for salt is the basis of another common name, Saltwort. These fleshy plants are somewhat translucent, suggesting that the common name Glasswort might refer to their appearance. The name, however, is derived from the ancient practice of burning these plants to obtain the minerals, collectively known as soda ash, necessary for making glass.

All species of Glassworts are easy to identify because of their thickened, somewhat brittle stems. The stems are fleshy until the end of the growing season, when they shrivel and present an entirely different appearance. In the fall several species turn a brilliant red.

The fleshy stems of all species are edible and tasty. In France they are considered a delicacy as a steamed vegetable, but they are virtually unknown for this use in the United States.

Collecting

Glassworts will grow only in saline areas. Along the Atlantic coast they often form very large stands and are easy to collect at low tide. Snap off the branches at least an inch above the soil because the bottom part of the stem can be quite woody.

Glassworts (Salicornia species): *(a) European Glasswort (a confusing common name since this species is native to North America); (b) Dwarf Saltwort; (c) the pudgy stems of this species are easy to collect and prepare.*

Salt and Vinegar Strips

30 Glasswort stems with the tough lower portion removed

5 large, tender Curly Dock leaves

Wrap the Glasswort stems in the Curly Dock leaves. Steam for 10 minutes. Serve cold. If the Curly Dock leaves (which add the vinegar flavor) are not young and tender, steam or blanch them before using.

Kudzu (Pueraria lobata): *a dense stand of Kudzu along a river bank.*

Kudzu has compound leaves with three leaflets.

KUDZU

KUDZU (*Pueraria lobata*) is a herbaceous vine in the pea family (Fabaceae). This exceedingly aggressive, non-native vine can grow more than 1 foot per day. Kudzu forms dense blankets over the foliage of native trees, making it easy to spot and easy to collect; it is usually found in dense patches. In Japan, the tuberous, starchy root is eaten, and Kudzu has a long history in Asia as a plant with medicinal uses. Kudzu is native to Japan and China and was brought to the United States in the late 1800s as an ornamental. Later, the U.S. Department of Agriculture used Kudzu for erosion control. In more recent times, the same agency has classified Kudzu as a noxious weed because of its ability to smother vegetation.

Kudzu spreads by vegetative runners and by rhizomes. It also spreads by seeds, which are contained in pods and mature in the autumn. This high-climbing, hairy vine has purple, pea-like flowers in dense clusters at the base of the leaves. The flowers are very fragrant and smell like grapes. The compound leaves of the plant have three broad, oval leaflets and curling tendrils.

Crunchy Kudzu Leaf Chips

15–20 Kudzu leaves

1 tsp. sugar

1 egg

1 cup very cold water (improves crispness of leaves)

1 cup all-purpose flour (or Curly Dock flour, see page 48)

3 tsp. olive oil

Wash leaves in cold water. Prepare a skillet with 3 tsp. of olive oil and heat on high. Beat 1 egg in a bowl. Pour in 1 cup of very cold water and mix until well blended. Be careful not to overmix the batter, as it may become too watery. Lay each leaf of three leaflets in the batter and then lay in the skillet with the hot olive oil. Turn each until golden brown and then lay the fritters on paper towels to drain excess oil.

Fried Kudzu leaves are great because they are one of the few mature leafy plants that are not fibrous. Of the wild plants we have tried, we think Kudzu is the best for frying.

Batter the Kudzu leaves thoroughly and gently fry them.

Kudzu leaves, flowers, and roots are edible. The root can be cut into pieces and crushed in water to release the starch. The young leaves can be fried in batter to yield crispy, tasty fritters. Pick and eat as much of this plant as possible; you are helping native plants!

Collecting

If you live on the East Coast of the United States, especially in the South, Kudzu can be found virtually anywhere, usually along roadsides and field margins. Collect 15–20 Kudzu leaves and place them in a 1-gallon plastic bag. Try to keep the leaves flattened in the bag for ease in frying.

Stinging Nettle (Urtica dioica): *(a) the stage when Stinging Nettle should be collected for use as a potherb; (b) the effective stinging hairs.*

MANY PEOPLE ARE SURPRISED to learn that a dermatitis-causing plant can be used in a multitude of ways and is one of the tastiest of all wild greens. Stinging Nettle (*Urtica dioica*) is a herbaceous perennial, 3 to 7 feet tall. The soft green, opposite leaves are 1 to 6 inches long, with sharply toothed margins. The entire plant is covered with stinging hairs. These hairs have a prestressed bulbous tip that easily breaks, forming a hypodermic needle that injects stinging chemicals such as acetylcholine, histamine, 5-hydroxy-tryptamine (5-HT, or serotonin), and possibly formic acid into the skin. This chemical cocktail causes a localized sting from which the species derives its common name; it is also called Burn Nettle, Burn Weed, and Burn Hazel. Stinging Nettle produces eruptions on the skin, a condition known as urticaria, which provides the genus name. The pain and itching from a Stinging Nettle encounter may last from only a few minutes to as long as a week. This is a much different reaction from that to Poison Ivy and is almost always much less severe.

Stinging Nettle is native to Eurasia and is now widely distributed in Canada and the United States, where it is found in every province and state except Hawaii. It is a nitrophile—that is, it prefers soils that have adequate nitrogen—and for this reason it is often found near habitations or where cattle are grazing. In addition to its comestible qualities, nettle makes a coarse but strong fiber and is still used in Siberia and Nepal to make cloth.

This armed plant is edible after the stinging hairs are neutralized, usually simply by boiling. It has a flavor similar to that of spinach when cooked and is reported to be nutritious. After Stinging Nettle flowers, the leaves develop gritty cellular inclusions called druses that can irritate the urinary tract. When mature, Stinging Nettle contains up to 25% protein, dry weight, which is high for a leafy green vegetable. The leaves and flowers can be dried and may then be used to make an infused beverage.

An alternative to Stinging Nettle is one of the false nettles (*Boehmeria cylindrica*), which is a common wetland plant throughout eastern North America.

Collecting

Stinging Nettle thrives in nitrogen-rich soil, so is often abundant around homes, at the margins of fertilized fields, and in barnyards, but it is also common in a variety of other disturbed habitats. Collect when the plants are less than 1 foot tall. This requires gloves. To determine whether the plants are tender, try snapping the stems. If this is difficult, the strong fibers have already started to develop.

Seeds of Stinging Nettle are also edible and can be collected in large quantities in late fall. Cut the stems and place them in a paper bag. As the plants dry the seeds will be released. These can be eaten raw or toasted, which we think improves the flavor. Place the seeds on a tray, put them in a toaster oven, and toast until the seeds begin to pop.

Stinging Nettle seeds mature in the autumn, when large quantities can be collected—using gloves, of course. Though tiny, the seeds are tasty.

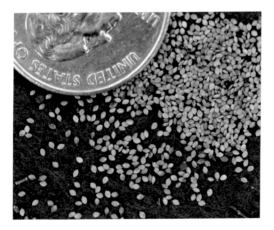

Nettle Omelette

2 eggs

1½ cups Stinging Nettle tender stems and leaves

¼ tsp. Field Garlic powder

cooking spray

salt and pepper to taste

In a large saucepan, boil ½ gallon water and drop the nettles in. After boiling for 15 minutes, drain and place nettles on paper towels to dry. When dry, cut and finely chop the nettle leaves and stems. In a bowl, beat the eggs, and stir in the chopped nettles. Season with salt and pepper. In a small skillet coated with cooking spray, cook the egg mixture over medium heat for about 3 minutes, until partially set. Flip with a spatula, and continue cooking 2 to 3 minutes. Reduce heat to low, and continue cooking 2 to 3 minutes, or to desired doneness.

BLACK WALNUT *(Juglans nigra)* is one of the best-known and most commonly utilized trees of eastern North America. The wood has long been favored for furniture and veneers. And the nuts are frequently collected in the late autumn for their rich flavor. Walnuts are readily identified by the large leaf divided into 7 or 9 leaflets. All parts of the plant have a distinct resinous odor. Unlike any other tree in North America, the center of the young stems has a chocolate brown color and numerous, narrowly spaced partitions. Black Walnut is therefore easy to recognize by its leaves and stems. The use of Black Walnuts is well known; they can be purchased in any grocery store and usually end up in cakes and cookies. We present a unique food that utilizes the distinct taste of this native nut.

RECIPE

Walnut Wild Plant Chips

Virtually any edible tender leaf can be used to make these distinctive snacks. We have selected four common plants that represent a diversity of tastes and textures (pine buds, Basswood leaves, Red Sorrel leaves, Dandelion leaves). If you experiment, make sure you select a leaf or bud that the walnut paste will adhere to.

1 cup Black Walnuts

water

salt to taste

Put the walnuts into a blender and add water until the consistency of humus. Coat leaves or buds with paste. Using a food dehydrator, dry for 12 or more hours. These chips can be frozen, and any surplus leaves and buds can also be frozen.

Black Walnut (Juglans nigra): *Black Walnuts in early autumn.*

Collecting

Black Walnut occurs through most of the eastern half of North America and, since it has been widely planted, it is often difficult to tell whether the trees are native or cultivated. Trees were often planted on farmsteads to ensure a handy supply of the nuts; they are also frequently planted along roads. So Black Walnut is an easy tree to find.

The nuts are mature when they start to fall from the tree. The outside of the fruit is green. A spongy (when fresh), thick, black layer encases the hard nut. Cleaning the nuts of the husk will stain your hands, so use gloves. Once all the green or black material is removed, the nuts can be washed but need to be quickly dried. Now comes the hard part. Black walnuts are tough nuts to crack. You will need a regular size hammer and a hard surface on which to place the nut. It takes some practice to find the right force for cracking the shell without smashing the nutmeats.

Removing the nutmeat is tedious work well suited for long winter evenings. Black Walnut contains large amounts of oils that can become rancid by the end of the winter, so refrigerated storage is best.

WHITE PINE (*Pinus strobus*) is the most desirable pine because of the presence of sugars, although any pine with needles that are flexible when young can be used. This includes Red Pine (*P. resinosa*), Longleaf Pine (*P. palustris*), and Loblolly Pine (*P. taeda*). Collect the young shoots in the spring just after the male cones (the pollen-bearing cones) have withered. The shoots should be flexible, with the stem easily snapped when bent.

BASSWOOD (*Tilia americana*) is a common deciduous tree throughout much of the eastern half of the United States and is readily identified by its edible broad leaves with heart-shaped leaf bases. Unlike most forest trees, Basswood usually has multiple trunks. Collect the leaves before all the leaves on a twig have developed; this will ensure that they are tender.

White Pine (Pinus strobus): *(a) White Pine buds at stage for collecting; (b) walnut paste–encrusted buds.*

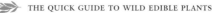

(a) Basswood tree (Tilia americana) *with characteristic multiple trunks; (b) Basswood leaves ready for harvest; (c) dehydrated leaves with paste.*

Red Sorrel (Rumex acetosella) *leaves: Red Sorrel is a common lawn weed with distinctive arrow-shaped leaves and a pleasant sour taste. Red Sorrel leaves are best early in the season, but in the Northeast they are available all summer.*

RED SORREL

THIS COMMON INTRODUCED WEED, Red Sorrel (*Rumex acetosella*), is also known as Sheep Sorrel and Sour Weed; the latter is an appropriate description of the pleasantly sour taste of the leaves. In northern states Red Sorrel can be collected throughout the summer. In warmer regions it must be collected in the late winter or early spring before it fruits and dies.

DANDELION

EVERYONE KNOWS DANDELION (*Taraxacum officinale*), usually from first-hand experience trying to expunge them from a lawn. Dandelion leaves can be harvested any time of the year. Choose the larger leaves because they are easier to coat.

 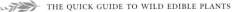

Dandelion (Taraxacum officinale): *(a) dehydrated*

Starches

SERIOUS WILD FOOD FORAGING requires the collection of starchy plants to provide the carbohydrates essential in our diet. Wild plant starches are the meat and potatoes for the wild food forager. For this reason some of the best-known edible wild plants have abundant starch and have been widely utilized. Some favorites are Groundnut and Arrowhead, and we also include lesser-known sources of carbohydrates.

AMERICAN LOTUS

AMERICAN LOTUS (*Nelumbo lutea*) is the most spectacular of all our native water lilies, with large, yellow flowers the size of dinner plates and distinctive fruits. Immense round leaves up to three feet in diameter die back in the winter to the surprisingly thin rhizomes. For the wild food forager, American Lotus is a great find because different parts of the plant are not only edible but also tasty.

This empress of ponds, rivers, and quiet waters is widespread in eastern North America. Although indigenous, it has been a problem in reservoirs, where it can form immense populations covering acres of water and interfering with hydroelectric processes and recreation. A benefit, however, is that lotus leaves provide important cover for spawning fish.

There are reports that the leaves are edible, but we have not eaten them. Better known are parts of the root system. Both species of Lotus spread by thin rhizomes. At the end of the season, swollen structures develop at the tips of the rhizomes; these are known as turions. Turions can break off the parent plant and float away to a new location to start a new population. Rhizomes and turions are discussed below.

The seeds are delicious and easy to collect. However, they remain palatable for only a short time before becoming indurate (hardened) and difficult to crack. Native Americans stored these seeds and ground them into flour. We have not tried this. Lotus seeds are the Methuselah of the seed

American Lotus (Nelumbo lutea): *the massive leaves of American Lotus have a shallow funnel shape. White flowers appear in mid- to late summer.*

world, with seeds known to have germinated after almost two thousand years. In our opinion, eating the fresh, tender seeds is the best way to use these tasty nuts.

Two species of lotus occur worldwide—the American Lotus and the Sacred Lotus (*Nelumbo nucifera*), which is widely grown in Asia for food as well as for its religious significance. Sacred Lotus has been introduced to the United States and is found in about a dozen states east of the Mississippi. It has large pink flowers but in other aspects resembles the American Lotus. In terms of wild food foraging, both provide edible seeds and underground parts. Sacred Lotus is cultivated for its rhizomes and turions, which are staple grocery items in most markets in Southeast Asia, where they are sold fresh or canned.

Collecting

Harvesting the rhizomes and turions is a muddy job. Collect them at the end of the growing season or shortly after frost, when it is easier to move among the plants. With your hand, follow the leaf stalk to the base of the

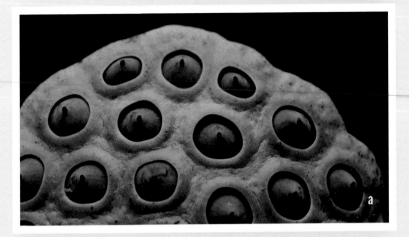

(a) The mature fruit of American Lotus. No other native plant has a fruit like this—a funnel-shaped, spongy structure with embedded seeds, the whole resembling a shower head. (b) Three stages of seeds: On the left are the extremely hard, mature seeds. Those on the right are immature, and those in the middle are at the best stage for eating. Within a week to ten days, the seeds harden.

Sacred Lotus (Nelumbo nucifera): *unlike the American Lotus, Sacred Lotus has pink flowers.*

~~~~9 RECICE ~~~~

Lotus Chips

The flavor of the turions and rhizomes is unremarkable, but

their interesting pattern makes them attractive. A simple way to prepare them is to cut the rhizomes and turions to make "wheels" ♠ inch wide. Fry these in vegetable oil until crisp, taking care not to burn.

(a) The Lotus rhizome, with the tip developing into two turions; (b) the prominent air chambers in the turions and rhizomes; (c) turions that are about 10 inches long and 2½ inches in diameter; the rhizome is 1 inch in diameter.

plant, where it is attached to the rhizome. Rhizomes can simply be pulled out of the mud. For turions, follow the rhizome until you find the swollen structure. A large number of turions can be collected in a short time.

ARROWHEAD

ANYONE SERIOUSLY INTERESTED in finding a dependable, tasty starch source needs to know this freshwater wetland plant found all across the continent. Not only does it form large populations, Arrowhead (*Sagittaria latifolia*), also known as duck potato, is easy to identify and harvest and simple to prepare.

Arrowhead grows in water and forms impressive stands along large rivers, like the backwaters of the Mississippi, where there are acres of plants. It is also abundant in lakes and marshes. Less frequently it grows in quietly flowing water. Arrowhead is a characteristic plant of diverse wetlands across the United States.

As both the common and Latin names indicate, the leaf is arrow-shaped. The width of the leaf can vary dramatically, from one foot across to as narrow as a half inch. Leaves are seldom more than 1½ feet long and arise from a short underground stem. Numerous other arrow-shaped leaves grow in the same wetlands as Arrowhead, but Arrowhead is the only one with milky juice, which is evident when one tears a leaf. Flowers are white and showy, and the spurred fruits are produced in the fall.

Like many wetland and aquatic plants, Arrowhead reproduces asexually by specialized underground structures called turions. The turion is a baby plant with a bottle—an ample supply of starch that can be used in the spring. The egg-shaped turions are formed in the autumn about when the leaves are dying back. The plants apparently live for only one year and then die back to the turions.

Arrowhead (Sagittaria latifolia): *(a) a dense stand of Arrowhead in late summer; (b) a plant in late autumn, showing the turions.*

Arrowhead Faux Bay Leaf

24 Arrowhead turions, cleaned

3 Swamp Bay leaves, dried

Boil the Swamp Bay leaves for 15 minutes, then add the Arrowhead. Reduce to a gentle boil for 10 minutes or until the turions are tender.

Collecting

Locating populations of Arrowhead is easy, since it grows in a great diversity of freshwater habitats, usually in full sun. Harvesting the turions is muddy work but well worth it! These storage organs are most efficiently collected by hand. Simply feel the base of the plant to locate the turions. It is possible to collect several dozen turions in a relatively short time. They can be baked or boiled and eaten like potatoes, but they have a nutlike flavor and a touch of bitterness. They are one of the tastiest of all wild plant starches.

GROUNDNUT

GROUNDNUT (*Apios americana*) is one of the few native plants of North America that has seriously been considered for improvement as a crop. All parts of the plant are edible, and, being a legume, Groundnut is nutritious as well.

A climbing vine that dies back to the ground each winter, Groundnut is a common wetland plant that favors the open margins of streams and lakes. It can climb up to 20 feet in one season. Leaves are divided into three or five segments and are about 5 inches long. In midsummer the attractive reddish brown flowers appear in compact clusters. The tips of the stems with young leaves can be eaten as a steamed vegetable.

(LEFT) *Groundnut* (Apios americana) *vines in flower;* (RIGHT) *Groundnut seeds can be eaten when green. The ripe, dry seeds can be stored.*

By late summer the long, narrow, green pods develop, and by autumn the pods open to release the hard, brown seeds. The young pods are tender and can be cooked and eaten like green beans. There is only a small window of opportunity for collection, however, as the pods soon become tough and stringy. In autumn the pods split open to reveal hard, brown seeds. The tubers are full of starch. In short, Groundnut is a vegetable store on a vine.

In winter the plant dies back to the numerous tubers at its base. Groundnut tubers are starchy, crisp, and tasty. The tubers are an exceptionally rich source of starch and can be collected in large numbers. Late fall or early winter, when the food is stored for the initiation of growth in the spring, is the best time to collect the tubers. Plants need to be located while the vines are still evident so the tubers can be found. They are usually located within the first foot of the soil.

Groundnut dries easily. For drying, cut the tubers crosswise into discs. These store well and can be added to soups and casseroles.

Groundnut Filé

1 lb of tubers

1 Tbsp. filé

Salt to taste

Clean the tubers. They can be peeled, but this is not necessary. Remove the rhizomes that connect the tubers, since they are tough and stringy. Place the filé into four quarts of water and boil for 15 minutes, then add the Groundnut tubers. Boil until tender, about 10 minutes.

(ABOVE) *Dehydrated Groundnut tubers;* (RIGHT) *Groundnut tubers are produced along a rhizome at regular intervals.*

TUBERS OF NUT SEDGE (*Cyperus esculentus*) have been found in several ancient tombs in Egypt, where they were valued as food. This favor is a contrast to the opinion of present-day farmers and gardeners, for whom Nut Sedge is a serious weed. In fact, it has ranked near the top of the ten most serious weeds in the world.

Native to the warmer regions of the Old World, Nut Sedge, also known as Tiger Nut and Chufa, among many common names, has been cultivated for millennia for the tubers, which are rich in oil and vitamins. It is frequently found in West African markets, as well as those in parts of western Asia, and is used as a health drink in southeastern Spain. But because of its weediness, in North America it is considered more of a bane than a blessing.

Nut Sedge looks like a grass with long, narrow leaves. Its stems are triangular in cross section rather than round, as in grasses. It grows about 2 feet tall and produces masses of tiny seeds.

Nut Sedge (Cyperus esculentus): *In late summer and early autumn, Nut Sedge produces characteristic seed heads. This is the best time to collect the tubers.*

The habitat of Nut Sedge is usually agricultural fields, where it is a fierce competitor with soybeans and other row crops. In these places it is often abundant, but being in an agricultural field and being a pernicious weed means it has probably been sprayed with an herbicide. Therefore, caution and care are needed when determining where to collect Nut Sedge. The plants must be uprooted to find the numerous tubers clustered at the base of the stem. Remove the tubers and discard the rest of the plant.

RECIPE

Field Garlic Chufa Nuts

1 cup Nut Sedge tubers

2 tsp. minced dried Field Garlic

Boil the tubers until tender, and allow to dry. Roll the tubers in a mixture of olive oil, salt, and minced garlic. Serve warm.

Nut Sedge tubers: These have been washed but not cleaned of the many fine (and tough!) roots attached to them.

OAKS (species of *Quercus*) include the best-known and most diverse group of trees in the eastern United States. The fruits of these well-known trees are acorns. All oaks have edible—though not always tasty—fruits.

These familiar trees can be divided into two general groups, the White Oaks and the Red Oaks, with clearly distinguishing characteristics. In White Oaks, the upper bark is rough, with obvious, distinct strips; in Red Oaks the upper bark is smooth. White Oak leaves never have prominent bristles at their tips, whereas most Red Oak leaves do.

Rock Chestnut Oak (Quercus prinus) *(a White Oak): White Oaks lack bristle tips on their leaves.*

White Oak acorns mature in one year, but with Red Oaks it takes two years for the acorns to ripen. This is obvious in Red Oaks because you can see mature and developing acorns on the same branch. It is important to learn the features of each group because of the quality of the fruits, the acorns. Red Oak acorns are often extremely bitter, so much so as to be virtually unpalatable. Usually much less bitter are acorns from White Oaks. In fact, some White Oaks have acorns that are delicious when fresh. The acorn is surrounded by a tough coat. This layer must be removed for most uses of acorns.

The acorn is particularly valued in the cuisines of some North American indigenous peoples and in many places in Europe and western Asia.

In North America tannins were often leached from acorns by putting the shelled nuts in a mesh sack and letting the sack sit in the waters of a fast-running stream for a week or so. Since most of us do not live near a clear, fast-moving stream, we recommend boiling the acorns repeatedly until most of the tannins are removed. The acorns can then be roasted just as are other tree nuts.

Again, all types of acorns throughout North America are edible. Raw acorns consumed in large quantities over time, however, may cause kidney damage. Although the acorns of White Oaks are more naturally sweet than those of Red Oaks, wild plants differ from cultivated plants in being much more variable. Some acorns from White Oaks may be more bitter than others, even when found in the same area. We strongly recommend staying with the White Oaks, especially the large acorns of the Rock Chestnut Oak (*Quercus prinus*), Bur Oak (*Q. macrocarpa*), Swamp White Oak (*Q. bicolor*), White Oak (*Q. alba*), and Swamp Oak (*Q. michauxii*). There are others; these are the ones we have used.

Collecting

In many regions of the United States, oaks are a conspicuous component of the vegetation. Keep an eye on the trees from which you plan to collect acorns. The falling of first acorns will be your indication that they are ripening. If possible, collect directly from the tree. If you must pick them up from the ground, look carefully for any fungal growth or acorn grubs.

Acorns, with a little work, can be processed into nutritious flour. A fist-sized flat rock or hammer works great as a nutcracker. After shelling the acorn, crush the meats into smaller pieces for boiling. Peel away as much as possible of the brown fibrous material (technically known as the seed coat) between the shell and the acorn meat.

Place the shelled, crushed acorn meats into a pot of boiling water. As the acorns boil, the water will become discolored. When the water is dark brown (after 15 minutes or so of boiling), strain out the acorn meats and switch them to another pot of boiling water. Continue this process until the nutmeats no longer taste bitter. We generally do three, four, or more water changes. The amount of boiling will vary depending on your acorns and your patience. Switching the acorns from boiling water to cold water seems to lock in the bitterness.

Crushing acorns with hammer.

Rappahannock Acorn Cakes

1 quart shelled acorns

1 cup cornmeal

¼ cup honey

Boil the acorn meats in water in a large gallon pot. Do at least three boilings until the water is only slightly discolored. Remove as much tannin as possible (repeated boiling). Drain and dry the acorn meats. Put the boiled acorn meats in a pot and mash with a potato masher. Lay mashed acorn meats onto a cookie sheet lined with aluminum and bake in an oven at 200°F for 1 hour. Crush or grind the dried acorn meats into flour (a coffee grinder works great). When acorns are ground fine, mix with corn meal. Add the honey. Mix the ingredients with enough warm water to make a moist—not sticky—dough.

Shape 4–6 patties of moist acorn meal and place on a cookie sheet lined with aluminum foil. Bake in the oven for 40 minutes at 350°F until slightly brown. Bake on each side, if necessary. Tasty acorn cakes!

Mok

Mok is a traditional Korean dish, highly valued and served on special occasions. As noted, Red Oak acorns have a repelling bitterness, but Red Oak acorns can be used in preparing *mok* because the preparation removes almost all of the bitterness. While we find nothing very special about the taste, this acorn product is unique and can be served with various accoutrements. It makes a real conversation piece at a dinner party—Got acorns?

2 cups shelled and chopped acorns

After shelling, soak acorns for two weeks, changing the water every other day. Using enough fresh water to cover the acorns, grind in a food processor until the acorn meat is a coarse slurry. Pour the slurry through cheesecloth, discard the solids, and pour the liquid into a quart saucepan. Heat to boiling with constant stirring. When the liquid thickens so that it barely drops from a spoon, pour into a container to cool. After jelling to the consistency of custard, the *mok* can be cut into squares. Garnish with black walnuts or chopped sorrel leaves. The coarse acorn flour can be stored or, for long storage, frozen.

(Clockwise from top left): fresh acorns, shelled and sectioned acorns, dried acorn flour, mok.

THE GENUS *Scirpus* includes a large number of aquatic, grasslike species in the sedge family (*Cyperaceae*) with common names such as Club-rush or Bulrush. The Softstem Bulrush (*Scirpus validus*) is a perennial with a creeping, horizontal rhizome that is edible. The round stems grow to 6 feet tall, and no typical leaves are present. Flowers are small and inconspicuous and produce hard, grainlike fruits in the late summer and autumn.

Scirpus validus, also known by the synonym *Schoenoplectus tabernaemontani*, is a wetland plant found in all fifty states as well as many other countries. It often forms large, dense stands in freshwater marshes, lakes, and other wet areas and is an important food for wildlife. It is also food for humans and has been utilized for its starchy rhizomes and grains.

Collecting

It is best to collect the edible rhizome in mid-fall. Look for a tall culm, or round stem, at the edge of the marsh; it will have drooping brown spikelets. Paddle up to the plant along the marsh or wade across the mud flat to find its rhizome, which is buried beneath the muck in the mud flat at a depth of less than one foot.

RECIPE

Softstem Bulrush

1 cup of the terminal ends of the rhizome

3 cups water (optional—3 cups beef or chicken broth)

¼ tsp. salt

Soak the terminal ends of the rhizomes in water overnight to loosen dirt. After an overnight soaking, gently clean the rhizomes with a small brush or potato peeler. Bring 3 cups of water to a boil in a saucepan, and then add the clean rhizomes of the bulrush. Add the salt and boil for 20 minutes. Remove the rhizomes and let them drain on paper towels; serve while hot. The boiled bulrush has a taste similar to that of boiled turnips.

Softstem Bulrush (Scirpus validus): (TOP) *a dense stand of Softstem Bulrush with developing fruit;* (BOTTOM) *the individual flower clusters characteristic of this species.*

SPRING BEAUTY (*Claytonia virginica*) is a small herbaceous plant that appears in dense stands in the early spring. The scientific name honors colonial Virginia botanist John Clayton (1694–1773). Spring Beauty is in the purslane family (Portulacaceae) and is native to eastern North America in moist woods and clearings. This small, slender plant has five-petaled, pink or white flowers that are ½ to ¾ inches wide and veined with darker pink. The stem bears only a single pair of narrow, succulent leaves. The flowers are produced in the early spring on branches 8 to 10 inches long, each with up to 15 flowers. After flowering, the plant forms a small capsule that contains numerous, tiny black seeds.

This small, vernal bloomer heralds the arrival of spring. Where you may find one flower growing there are likely to be many, sometimes in the hundreds and thousands. Within a week or two, though, the plant disappears, and you will not find it, no matter how hard you look.

The habitat of Spring Beauty is rich, moist soils, wooded floodplains, and sometimes wet meadows. Like many other spring ephemerals of the deciduous forest, it completes its life cycle in only a few weeks, drawing on the reserves stored in the tuber at the base of the plant. The edible, underground tubers were relished by Native Americans for their chestnut-like flavor. They contain an ample store of carbohydrates and, since Spring Beauty can grow in dense stands, it can be a profitable source of calories. This delicious tuber can vary in size from that of a peanut to as large as a bulb of garlic. The tuber can be eaten fresh or prepared by boiling, roasting, or frying.

Collecting

Collecting Spring Beauty tubers requires patience and a small digging tool such as a large spoon or a similar instrument. Collect the tuber only if this plant is abundant so as not to reduce the local population of this delicate flower. Harvesting is a tedious job; be prepared to get dirty because it is more comfortable to lie down for extended periods of time collecting tubers.

(LEFT) *Spring Beauty*
(Claytonia virginica):
the flowers of Spring Beauty
are white with pink veins.
(RIGHT) *Tubers of Spring*
Beauty: the darker and
larger tubers are older
and can give rise to
several flowering stems.

Sauteed Spring Beauty Tubers

½ lb cleaned Spring Beauty tubers

3 tsp. olive oil

Carefully clean the tubers under running water. Washing the dirt
and grit out of the tubers is very important; you don't want to
grind your teeth on any remaining grit. Soaking in warm water
for half an hour will help remove the grit from the outer layer
of the tubers. Prepare a frying pan with the olive oil and heat on
high. Lay the cleaned tubers in the frying pan with hot olive oil.
Turn each tuber until lightly roasted brown. After 5 minutes or
so, place the tubers on paper towels to absorb excess oil. Serve
hot. Delicious!

Grains and Plants Used Like Grains

WE ARE A NATION OF GRAIN EATERS. Grain is the technical term for the one-seeded fruit of members of the grass family. Each day most of us eat wheat or corn in one form or another. Corn provides oil, starch, and a host of other products that make it the basis of our food security. Wheat and barley, on the other hand, are the staff of life in much of the rest of the world. They were selected from among the native grasses in Mesopotamia because of their large grains and because they exhibit tillering, the ability to produce additional stems from the base of the plant, essentially providing several fruiting stems from one grain. The most widely utilized grain on a global scale is rice, not to be confused with our native Wild Rice (*Zizania palustris*). In North America only a few native grains other than corn were cultivated, although several were harvested in the wild, most notably Wild Rice.

Many native grains are edible, though the grains of Perennial Ryegrass (*Lolium temulentum*) can be toxic because of the presence of a micro-organism. We are taking the liberty of including plants that are not true grasses but have grainlike fruits or seeds that are eaten like true grains.

CANE

CANE (*Arundinaria tecta* and other species) is a true bamboo, that is, a woody member of the grass family. The height of the plant varies considerably; it can be up to 8 feet tall and is sparsely branched except when flowering and fruiting. Leaves are up to 1½ feet long and narrow, about 2 inches wide. Most populations retain their leaves through the winter.

Because of its rigorous rhizomes, Cane produces large monocultures known as canebreaks. This wetland community is one of the rarest in the eastern United States, and few intact examples remain. Cane once formed extensive stands throughout its range in the southeastern United States, with some populations as far north as New York. Like many of the native plants of the American Southeast, Cane is fire-adapted to burning or other activities that mimic fire, like vegetation removal, which stimulates the plant. There is now a concerted effort to restore these communities and learn how to grow Cane for wetland restoration.

Botanists do not agree on how many species occur in eastern North America, so Cane, also known as Switch Cane, may be treated as a subspecies of *Arundinaria gigantea*. A recently described new species from the southern Appalachians is known, appropriately, as *A. appalachiana*. All three species are similar and are separated only on technical characteristics. For our purposes, they can be considered together because of their similar use and behavior.

The behavior of bamboos is puzzling to scientists because what triggers flowering and fruit production is unknown. Like all bamboos, Cane fruits once and then dies. This fruiting is completely unpredictable, and locating a fruiting population of Cane, unfortunately for the wild food enthusiast, is a very special occasion. We have found fruiting Cane only three times in our combined 75 years of field work.

Coming upon a fruiting stand of Cane is a noteworthy occasion, especially if the grains have not been harvested by wildlife, which also eat the

Cane (Arundinaria tecta): *A road in the Great Dismal Swamp National Wildlife Refuge in Virginia is lined with miles of fruiting Cane in May. As the fruits mature, the plant goes into senescence (grows old) and turns brown.*

A fruiting branch of Cane with mature grains.

Ripe grains of Cane: note the size—cane has the the largest grain of any of our native grasses.

shoots and leaves. While an outstanding food for wildlife, the grains of Cane have not featured prominently in human diets because of its erratic fruit production.

Collecting

Look for Cane at the margin of forested wetlands and savannas, where it often forms large populations. The flowers are produced in late winter and, like those of other grasses, are underwhelming. Unlike many grasses, the appearance of flowers does not mean there will be fruiting. Fruits, if produced, will appear in late April to mid-May.

The fecundity of Cane has to be seen to be believed. The branches are loaded with grains, the largest grains of any native grass in the eastern United States—up to ½ inch long.

Harvest the grains by bending the fruiting stems over a basket or flat pan—a pizza pan works well. Wear gloves because you will be stripping off the grains and the broken stems can be sharp. It is possible to collect several pounds in a half hour.

The grains are healthy and delicious when fresh, with a chewy, nutty taste, but they need to be dried for preservation. Like other wild grains, removing all the tiny attendant parts—collectively known as chaff when it is removed—is a real pain. Chaff contains silica in its cells that, while not toxic, is less digestible than the grains. If you want to eat the grains fresh, the easiest way is to rub the grains between your hands. However, to prepare larger quantities requires roasting the grain.

Roasted Cane grains: processing breaks some of the grains; the white grains are the broken ones.

Place the fresh grain on a cookie sheet and bake uncovered at 200°F for half an hour. After cooling, place the grains into a food blender ½ cup at a time. Use the pulse setting to minimize breakage. Remove as much chaff as possible by pouring from one container to another with a gentle breeze. Then, put the grain in about a quart of water and agitate by hand. The grains will sink; remove the floating chaff and repeat the process. About six cups of roasted grain will yield approximately ½ cup of clean grain. It is virtually impossible to remove all the chaff. Dry the cleaned grain at room temperature. It can be stored for a long time in this condition, and freezing will extend storage time.

Cane Crispies

1 cup roasted grain

1 Tbsp. vegetable oil (buffalo fat works great but is hard to get!)

salt to taste

Grind the grain in a food processor to make a coarse flour. Mix the flour with enough water to form a small patty about 3 inches in diameter. Fry until crisp, taking care not to burn. These are chewy and tasty. Both the flour and the crispies can be frozen.

MANNA GRASS

MANNA GRASSES (*Glyceria melicaria* and other species) are wetland plants with several species in North America. One of the most common is *Glyceria melicaria*, which often forms large, uniform stands at the margins of lakes, bogs, and streams in the northeastern part of the country, where it is also known by the uninspired common name of Northeast Manna Grass.

All manna grasses prefer wetland habitats and are characterized by having the leaf sheath (the part of the grass leaf that wraps around the stem) without a slit—unlike almost all other grasses. The grains are small and ripen in midsummer on drooping branches. Although small, most of the tiny flowers produce a grain, so that a stand of Manna Grass is a welcome sight for a wild foods forager.

Collecting

Two of the attractive features of this grass are the abundance of grains and the ease in harvesting them. In addition, removal of the chaff is simple, and it is possible to obtain grain relatively free of any chaff.

Cut the fruiting stems and allow them to dry for a day. Then strip off the grains. Winnow if necessary. While they can be eaten immediately after collection, with a fresh, woodsy taste, the grains are even better if roasted. Roast at 200°F for 15 minutes or until brown, taking care not to burn.

Manna Grass (Glyceria melicaria): with ripe grains in midsummer.

Red Sorrel Pilaf

1 cup fresh Red Sorrel leaves

1 cup cleaned Manna Grass grains

2 cups water

salt to taste

Bring the Red Sorrel leaves to a boil. Reduce heat and add roasted grains. After roasting, boil the grains until tender. This takes about 15 minutes.

Manna grass: (a) roasted grains; (b) cleaned, fresh grains.

THE COMMON NAME RIVER OATS (*Chasmanthium latifolium*) derives from the usual habitat of this attractive grass along rivers and other watercourses in the eastern half of the United States. Often forming large populations, River Oats fruit late in the fall, preferring areas that have been scoured by seasonal flooding. Unlike many grasses, River Oats can thrive in partial shade, providing the source of another common name, Wood Oats. This native grass is no relation to true oats but does produce a tasty grain that can be used similarly to oats.

Plants are waist high when fruiting, with large brown grain heads contrasting with the dark green of the leaves. The leaves are wider than those of most grasses (up to 2 inches wide), hence the species name *latifolium*, which means broad leaf. But the large grain heads are deceiving because the grains are quite small and often only a few per head mature.

*River Oats (*Chasmanthium latifolium*): a fruiting stand of River Oats in late fall.*

(OPPOSITE) *A fruiting stalk of River Oats showing a single grain in the whole stalk.*

Threshed grains of River Oats.

River Oats can be found in riverine forests, especially on swales that are flooded at high water. The time of collection is important because this plant will drop the grains when they are fully ripe. This means keeping an eye on the population you want to harvest. Cut the fruiting stalks and put them in a large plastic bag. Remember, the fruiting heads have few grains, so large quantities need to be collected. Spread the stalks on newspaper in a dry, cool place. During this process the grains will separate from the fruiting heads. Beating the sheaves will remove any remaining grains.

The next step is extricating the grain from the tiny but tough scales that surround the grain. This is tedious and is best accomplished by rubbing the grains between your hands. Since the scales of River Oats are sharp, gloves are needed. Finally, washing the grains will remove any stray bits of chaff. Put the grain in a bowel of water and let it settle. The chaff will float to the top. Dry the grains immediately at room temperature.

If you started with three large garbage bags of sheaves, you will have about 2½ cups of grain, but the percentage of recovery will vary considerably depending on the plants, the season, and the locality.

River Oats Oatmeal

2 ½ cups River Oat grains

4 cups of water

Grind the grains in a coffee grinder until most of the grains are at least half the size of the originals. Boil until tender. The final product will be darker than oatmeal, with a stronger but pleasant flavor and abundant fiber.

River Oats Pasta

2 ½ cups River Oats

2 ½ cups all-purpose flour

salt to taste

Grind the River Oats as fine as possible in a coffee grinder. Mix the two flours with enough water to form a dough. Roll out the dough on a cutting board and cut into strips. Allow to dry and use as any other pasta. This pasta will be darker and have a distinct, pleasant flavor.

YELLOW POND LILY (*Nuphar lutea*) is one of the most common and widely distributed aquatic plants in the United States, growing in lakes, rivers, ponds, and freshwater tidal regions. The shape of the leaves varies from oval to elongate. These are borne on long leaf stalks that arise from a massive rootstock. Leaves can be up to two feet long, with a slit extending one-quarter the length of the leaf. Flowers appear from spring to frost on single stalks rising above the water, with bright yellow flowers about the size of a golf ball. The flowers produce the many-seeded fruit.

A boat is useful to collect a large number of the fruits. Check the color of the seeds to ensure that they are mature; they should be brown and hard. Remove the seeds, and wash to remove the slime. Fresh seeds can be roasted and ground or popped. To pop the seeds, place them in a toaster oven and toast using the "dark" setting.

Yellow Pond Lily leaves float on the surface with the flowers raised slightly above them. (OPPOSITE) *Mature fruits and seeds are available most of the year.*

Flowers

FLOWERS MAY SEEM LIKE an unusual source of food, yet such common foods as broccoli, cauliflower, and artichoke are flowers—though in the bud stage. Pollen, the male cells of the plant, is nutritious but is seldom available in quantities suitable for harvest. One of our most common wetland plants, Cattails, has been a source of pollen for food for millenia. The other flowers we have selected are common and easily prepared.

BLACK LOCUST

BLACK LOCUST (*Robinia pseudoacacia*) is a tree in the pea family (Fabaceae). It is native to the southeastern United States but has been widely planted and naturalized elsewhere in temperate North America, Europe, and Asia. It is considered an invasive species in some areas.

The bark is black and deeply furrowed. A fast-growing tree, Black Locust can reach a height of 100 feet. Leaves are alternate and compound, over 20 inches long, and with 9 to 19 oval leaflets, each about an inch wide. Each leaf usually has a pair of short, sharp spines at its base. The showy white flowers are fragrant and are borne in hanging clusters more than 8 inches long.

The flowers are the only edible part of the tree; there are reports of poisoning from ingestion of other parts of Black Locust. The flowers appear in early spring, April through May. They can be eaten raw, and they have a somewhat crunchy texture, sweet taste, and pleasant fragrance.

Collecting

Black Locust usually occurs in stands that are very easy to recognize when in flower. These trees are often found along fence rows and in disturbed areas.

Black Locust Fritters

4 flower bundles

1 egg

1 cup of very cold water (cold water keeps the flowers crisp)

1 cup all-purpose flour

3 tsp. olive oil

Gently brush away any insects on the flowers. Shake vigorously, but do not wash the flowers, as this removes the nectar that provides considerable flavor. Prepare a skillet with 3 tsp. of olive oil and heat on high. Beat 1 egg in a bowl. Pour in 1 cup of very cold water. Mix egg and cold water until well blended. Mix in 1 cup of all-purpose flour. Be careful not to overmix the batter, which would make it tough. Place each flower bundle in the batter and then into the hot oil. Turn until golden brown. Lay the fried flower bundles on paper towels to absorb excess oil. Voila! Blossom fries.

When flowering, the tree can easily be spotted by the abundance of white flower clusters that hang from virtually every branch. Collect four or five clusters and place them in a large plastic bag. Tie the bag to retain the fragrance.

(OPPOSITE) *Black Locust* (Robinia pseudoacacia): *the white, conspicuous flower bundles of Black Locust.*

Black Locust Yogurt Dip

Follow the instructions for Black Locust Fritters for cleaning the flowers.

2 cups Black Locust flowers (individual flowers do not have to be removed from the stem)

1 pint full-fat yogurt (full-fat yogurt absorbs more of the flavors of the flowers)

2 Tbsp. Sassafras filé (see page 30)

salt and pepper to taste

Thoroughly mix the Sassafras filé with the yogurt and refrigerate overnight. Add the Black Locust flowers and return them to the refrigerator for one day. Remove the flowers from the dip before serving to make dipping easier.

The flower bundles of Black Locust can be easily picked right off the tree and eaten raw.

Cattails (*Typha* species) are arguably the most widespread wetland plant in North America and certainly one of the best known. There are several species in the United States, but all are used similarly, as are their relatives in far-flung corners of the globe, where their use is well documented.

Cattails are the panacea for wild food foragers—readily identified, easily distinguished from any toxic or harmful plant, available in large numbers, useful anytime of the year, and with all parts of the plant edible. In addition, Cattails actually taste good! For these reasons Cattails are ideal edible plants for first-time foragers.

*Cattails (*Typha *species): a stand of Cattails with a flowering stem. The male flowers are at the top and are past flowering. The light green portion below are the mature female flowers, which are at the stage when they can be eaten. If any of the female stalk is brown, it is past the stage for collecting.*

Detail of female Cattail flowers: A good indication of the proper maturity for collecting is when the tiny white structures, the styles of the flowers, are evident.

Most wild food recipes use rhizomes or young leaves, both of which are excellent. We have selected two recipes using the female and male flowers. These are borne on the same stem, with the male portion above the female. They ripen at differing times, so plan your collecting accordingly.

FEMALE FLOWER STALKS

THE FEMALE FLOWERS have a pleasant, meaty flavor and texture. Flowers form a thin layer around the central stem and can easily be eaten like a corn dog on a stick. We have selected one simple recipe for this part of the plant.

Collecting

Collect the flowering stalks when the female flowers are at the proper stage. Leave at least 5 inches of the stalk below the female flowers to use as a handle. Cut the section above the female flowers to remove the male flowers. Garden clippers are necessary because of the tough stem.

RECIPE

Cattail Corn Dogs

10 female flower clusters, cleared of male flowers

1 cup cornmeal

salt to taste

½ cup vegetable oil

Heat oil in a large frying pan. While oil is heating, roll the female stalks in the cornmeal. Fry quickly until the cornmeal darkens, taking care not to burn. Cool on absorbent paper.

Mass of male Cattail flowers just opening in the center of the stalk: This is the ideal stage for collecting pollen.

MALE FLOWERS

Collecting

The window of opportunity for collecting male flowers is short, a week or less depending on weather. The edible part is the bright yellow pollen that is produced in immense quantities. The first evidence of mature pollen is usually in the middle of the male flower stalk.

Cut the male flower stalk just above the female flowers. Garden clippers are helpful, as the stems can be quite tough. Immediately place the male stalks in a paper (not plastic) bag. Then, place the stalks on paper in a warm, dry area to allow the pollen to be released. This will require about five days at room temperature.

To collect the pollen, place the dry stalks in a large container and rub the stalks between your hands. Fifty stalks will yield about three cups of material, which includes both the pollen and the remnants of the male flowers.

A sieve with the mesh size of a tea strainer works well to remove the flower parts. Removing all the flower parts is impossible, and these remnants will not interfere with the flavor. Two sievings will yield about ¼ cup pollen. This can be stored in a dry, cool place or frozen until used.

Cattail pollen ready for cooking.

Mesopotamian Pollen Paddies.

Mesopotamian Pollen Paddies

For millennia Arabs living in the marshes of Iraq have used cattails for food, mats, and floatation devices. This recipe is from an Iraqi colleague from the marshes.

½ cup Cattail pollen

Cut cheesecloth into 4- x 5-inch squares. Spoon 1 tsp. pollen onto the cheesecloth, and fold the cheesecloth over the pollen. Steam for one hour. A vegetable steamer works fine.

The flavor is pleasantly sweet with a nutlike aftertaste. The paddies can be frozen for later use. Serve Mesopotamian Pollen Paddies with coffee.

DAY LILIES ORIGINATED in East Asia and have been cultivated in North America for several hundred years as an ornamental. So successful was their introduction that they are now weedy and found in most of the lower 48 states and parts of Canada. The most widely distributed is *Hemerocallis fulva*. The name refers to the orange color of the flowers (tawny brown is *fulva* in Latin). These are long-lived perennials with strap-shaped leaves arising from a cluster of swollen roots. The large, familiar flowers appear in the spring, and each lasts for only one day. Orange Day Lilies spread by aggressive rhizomes and soon establish large patches along roads and in old gardens, cemeteries, and waste places.

Day Lilies have been eaten for perhaps thousands of years and are a component in Chinese cooking, especially the dried buds, which are used as a thickener for soup. The roots are also edible; they have a mild taste and a crunchy texture. In recent years, hundreds of new varieties have been developed; they include genetic material from species with unknown edibility. For that reason, we recommend collecting only the well-known and time-tested *Hemerocallis fulva*.

Orange Day Lily (Hemerocallis fulva): *a large population of Day Lilies along a road.*

Tuberous roots of Day Lily: note the white, firm flesh evident on the cut root.

Pickled Day Lily Buds

Select younger buds. While all stages of the buds are edible, those that do not show any of the characteristic orange color are best.

2 cups Day Lily buds

standard brine solution: 2 Tbsp. salt, 2 cups white vinegar

Bring the brine to a full boil. Place the buds into a sterilized canning jar and cover with the boiling brine. Tightly screw on the lid. Leave the buds for about a month.

(OPPOSITE) *Any size Day Lily bud is suitable for dehydration. Younger buds are preferred for pickling.*

Day Lilies are common around old homesteads and in cemeteries. They often spread to roadsides.

REDBUD

REDBUD (*Cercis canadensis*) is a well-known harbinger of spring throughout the eastern United States. This small tree produces masses of pink flowers before its leaves appear, garlanding roadsides and forest margins, its most common habitats, with pink mists. The small (about 1 inch long) flowers do not last long. By midsummer the distinct podlike fruits develop. Unlike the flowers, the fruits are not edible, and there are reports that other parts of the plant are slightly toxic. While not substantive in terms of calories, the flowers are a nice addition to a salad or as a topping on ice cream. The flavor is mild and sweet, and the texture is slightly crunchy.

To harvest the flowers it is best to cut the twigs and shake them vigorously to dislodge the small insects that feed on the nectar. Then remove the flowers by hand. Do not wash Redbud flowers, or they will turn into a soggy, tasteless mess.

Redbud (Cercis canadensis): *Unlike those of most trees, Redbud flowers emerge directly from the bark on the trunk of the tree. The easiest way to harvest them is by clipping short branches.*

Sweets

OBVIOUS SOURCES OF SWEETS are the numerous native berries and other fruits in the flora of North America. Because most of these are so well known, we have highlighted two that are seldom included in discussions of native sweets.

One of the best known and most widely used of all native sweeteners is maple syrup. We do not cover syrup production in this book. Those interested can consult several websites, including www.umext.maine.edu/onlinepubs/pdfpubs/7036.pdf.

INDIAN STRAWBERRY

INDIAN STRAWBERRY (*Duchesnea indica*) is a small, herbaceous perennial with leaves and fruit similar to those of a true strawberry (species of *Fragaria*), its relative in the rose family, or Rosaceae. Other common names of Indian Strawberry include Mock Strawberry and False Strawberry. It can easily be distinguished from true strawberries by its yellow flowers; all true strawberries have white or slightly pink flowers. The flowers of Indian Strawberry appear from April until frost. The leaves are light green and have three parts. Native to eastern and southern Asia, it was introduced as an ornamental and has naturalized in eastern North America.

Indian Strawberry prefers moist, well-drained soil in partial sun. This low-growing plant sends out runners, or stolons. New plants can develop at each node of the stolon.

There are few references to the edibility of Indian Strawberry, which is surprising because the fruits are tasty. The round ½-inch fruit of *Duchesnea* is a small red berry-like structure with a juicy, slightly sweet taste and a slightly gritty texture; it's entirely covered with small red seeds (technically achenes). These fruits are a source of sugar, protein, and ascorbic acid (vitamin C).

Indian Strawberry (Duchesnea indica): *the three-part leaves are very similar to those of true strawberry.*

Indian Strawberry fruits differ from those of true strawberries in that each individual seed has a fleshy covering.

Indian Strawberry Jam

1 cup Indian Strawberries

2 Tbsp. sugar

¼ cup water

Pour all of the fruits of the Indian Strawberry into a large bowl and rinse to make sure there are no insects or dirt. Next, mash the fruits thoroughly. Transfer the mashed fruit to a small bowl and stir in ¼ cup water and 2 Tbsp. sugar. Let the mixture sit overnight in the refrigerator. In the morning you will have fresh Indian Strawberry jam for your toast!

The fruits of Indian Strawberry are easy to find and collect. Usually, there are many plants growing together, with enough red fruits to prepare a delightful jam. In a snap you have a cup or more of the red fruits!

Collecting

You will need to look along the edge of the forest, usually near a field, to find several plants growing in close proximity. Indian Strawberry is also a common lawn weed, especially in slightly damp soil.

PAWPAW (*Asimina triloba*) is native to North America and is in a family of plants (Annonaceae) that includes other tropical edible plants such as Custard Apple (*Asimina reticulata*), Soursop (*Asimina muricata*), and Cherimoya (*Annona cherimola*). Several members of this family have bark, leaves, and roots used in folk medicines. Pharmaceutical research has found antifungal, bacteriostatic, and especially cytostatic capability of some chemical constituents of the leaves and bark. A large number of chemical compounds, including flavonoids, alkaloids, and acetogenins, have been extracted from the seeds and other parts of these plants.

Pawpaw is a deciduous tree that grows along streams and rivers in rich, moist alluvial soils in hardwood forests. It occurs in twenty-five states in the eastern United States ranging from northern Florida to southern Ontario (Canada) and as far west as eastern Nebraska. It grows in thickets because it spreads by underground stems. The red, succulent flowers emerge before the leaves in mid-spring; the blossoms occur singly on the previous year's wood and are up to 2 inches in diameter. Pollination is by flies and beetles, which is suggested by the appearance of the flower; it has dark, meat-colored petals and a fetid smell.

Pawpaw fruits are oblong to cylindrical berries that are typically ½ to 5½ inches long and ½ to 4 inches wide. Each fruit can weigh ¼ to 4 pounds. Fruits may be borne singly or in clusters that resemble the "hands" of the banana stalk. The shelf life of a tree-ripened fruit stored at room temperature is only 2 to 3 days. With refrigeration, Pawpaw fruit can last up to 3 weeks with good eating quality. Pawpaw has the largest edible fruit of any native tree in the eastern half of the United States. Its taste resembles a creamy mixture of banana, mango, and pineapple, and it has the consistency of custard. The fruit of Pawpaw never became commercially available because of its short shelf life. This may be changing, however, as it is being investigated for its commercial value as a specialty fruit. Pawpaw fruits are nutritious and high in vitamins and minerals, generally containing these nutrients in amounts similar to those found in bananas, oranges, and apples.

Pawpaw (Asimina triloba): *pawpaw flowers mature just as the leaves are beginning to appear.*

CAUTION

WITHIN THE FRUIT, there are two rows of large, brown, bean-shaped seeds, each about 1½ inches long. These seeds contain alkaloids in the endosperm that are emetic. If the seeds are chewed, seed poisons may impair mammalian digestion but, if swallowed whole, seeds may pass through the digestive tract intact. Except for the ripe fruit, all other parts of the plant should be considered toxic.

Collecting

Pawpaw fruits ripen in late September through mid-October. The fruits start to turn yellow, become splotched with brown, and have a soft but firm feel when lightly squeezed. The riper the fruit, the blacker it becomes and the uglier it looks. The fruit is completely ripe when the outer skin turns black. These blackened fruits can be collected from the ground if they are not damaged. Remove the outer skin and take out the seeds. Discard these seeds or plant them. When ripe, the flesh of the fruit can be scooped out with a tablespoon and eaten fresh or used in pudding and bread.

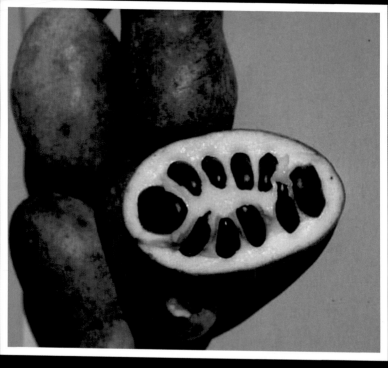

Mature Pawpaw fruits: the seeds are large and black.

Pawpaw Bread

2 cups all-purpose flour

2 eggs, beaten

3 cups mashed ripe Pawpaw fruits

Preheat a conventional oven to 350°F. Lightly grease a 9- x 5-inch loaf pan. Put the flour in a large bowl. In a separate bowl, stir together the beaten eggs and the mashed Pawpaw fruits until well blended, then add this mixture to the flour and stir until moist. Pour the batter into the prepared loaf pan. Bake in the preheated oven for 60 to 65 minutes. Let the bread cool in the pan for 10 minutes, and then turn out. Serve Pawpaw bread with chili or soup.

Cordials

A CORDIAL IS A SWEETENED alcoholic drink produced by infusion, that is, by soaking material in alcohol. Originally produced as a health tonic, cordials are now used as after-dinner drinks. The name derives from the Latin word for heart in reference to the putative effect of such drinks. We are not touting the health benefits of our cordial but rather the great taste.

A sweet, fruity cordial is an excellent way to finish a meal—especially a meal with some of the more wild-tasting plants we discuss. Virtually any fleshy fruit can be made into a cordial. We include a single example of what we considered the tastiest of the more than 60 different cordials we have tested.

BLUEBERRIES

MANY DIFFERENT FRUITS can be used to make cordials, as long as the fruits are fleshy and have aromatic components that infuse in alcohol. We have chosen blueberries (*Vaccinium* species) because they are so easily recognized and are widespread throughout eastern North America.

All native blueberries are suitable, but those that mature late in the season, such as Farkleberry (*Vaccinium arboreum*), are often drier and therefore less flavorful than the blueberries that mature in the summer, such as Highbush Blueberry (*Vaccinium corymbosum*).

Collecting

Locate a stand of wild blueberries and observe them at the beginning of the season to determine when the fruits might be produced. Fruits are produced in June in southern states and from late July until frost in northern states. Wash the berries after collecting, taking care not to break the skin. Let the berries dry at room temperature before preparing the cordial, but make sure they do not dry excessively.

Highbush Blueberry (Vaccinium corymbosum), *a common and widespread species that produces fruit in midsummer.*

Wild Blueberry Cordial

2 cups blueberries

½ cup sugar

1 cup vodka (unflavored)

Place the blueberries in a 4-cup container. A flat container is best because it allows more surface area of the berries to be exposed. Sprinkle the sugar evenly over the berries. Allow this to sit for one week at room temperature, checking for fungal contamination. During this time the sugar will draw out much of the moisture of the berries. After one week, stir the berry-sugar mixture and add the vodka. Leave for three months in a cool place but do not refrigerate. Strain the cordial into a stoppered bottle and store in a cool, dark place. The residual berries can be used in a fruit salad.

The delicious end product! The dark blue Wild Blueberry Cordial has a pleasant aroma and fruity flavor.

Mushrooms

MUSHROOMS ARE NOT PLANTS; they are in the Kingdom Fungi but were traditionally included in the study of plants because they have plantlike characteristics. We include them in this book because any wild food enthusiast will encounter mushrooms while foraging and also because there are several that are easy to identify. The features used to distinguish among mushrooms are entirely different from those used to identify plants. This should not discourage anyone interested in learning about these fascinating organisms. Rather, the novice should invest in a good book dealing with fleshy fungi (which include mushrooms, shelf fungi, puffballs, and more) and patiently pursue learning how to recognize local mushrooms.

This is not a book about mushrooms. Therefore, we have selected only a few widely distributed, easy to identify, and easy to prepare species.

Never ingest an unknown mushroom! Be certain that you are not allergic to mushrooms so that an allergy is not mistaken for poisoning. For emergency help in mushroom poisoning, call the Poison Hotline at 1-800-222-1222.

OYSTER MUSHROOM

WIDESPREAD, EAGERLY SOUGHT as food, and easily identified, the Oyster Mushroom (*Pleurotus ostreatus*) is named for the shape and gray color of the cap rather than for a resemblance to the taste of oysters. Its frequency and ease of recognition along with its typically prolific fruiting make it a prized food mushroom worldwide. Oyster mushrooms are now available, both dried and fresh, in food stores throughout the United States. But they are so common in nature and so easily identified that there is no need to get them from a store.

The host range of this mushroom is broad and includes many species of living, dying, and dead trees. Where does Oyster Mushroom grow? In the Middle Atlantic States most frequently in floodplain forests, but it could turn up just about anywhere, including in a neighborhood near you.

The season for Oyster Mushrooms is more specific than its host tree—cool weather, especially after a period of warm, humid weather. But it retains its delightful felicity in fruiting at unpredictable times.

Oyster Mushroom is a shelf fungus, which emerges at right angles to the trunk of the tree, typically producing troops of caps, sometimes in immense quantities. The top of the cap—slightly tacky to the touch when fresh—is gray or purplish gray, and the edge is curved; these curved caps help distinguish Oyster Mushroom from the many species of ubiquitous woody shelf fungi. Older mushrooms will have cracked caps with upturned rather than curved margins. Unlike the majority of shelf fungi, which are tough and can grow for years, the Oyster Mushroom decays within a week or so after appearing, leaving the tree with no trace of its once attractive parasite. Unlike most shelf fungi, Oyster Mushroom has gills rather than pores.

The large, widely spaced gills are on the undersurface of the cap. Oyster Mushroom produces dirty white to lilac-colored spores. To obtain a spore print, place the mushroom, gills down, on a piece of white paper in a warm,

Oyster Mushroom (Pleurotus ostreatus): a fruiting cluster of Oyster Mushroom on a hackberry tree. These mushrooms have just emerged within the past 48 hours and, because the weather has been cool, there are no insects. Cutting the cluster from the tree with a knife avoids breaking the caps.

dry place. Maximum sporulation will occur within four hours. The base of the mushroom, the stalk, is very short or entirely lacking and is usually hairy. With its curved rubbery caps and lilac spore print, the Oyster Mushroom is distinct from any other shelf mushroom. There are no toxic shelf fungi with which it can be easily confused.

The curved cap and widely spaced gills are characteristic of the widespread and often abundant Oyster Mushroom. The shape and color of the cap give the Oyster Mushroom its common name; its flavor does not resemble that of the shellfish.

A spore print from the Oyster Mushroom. No other of our shelf fungi has a sordid (dirty) lilac spore print.

Oyster Mushroom Soup

about 10 fresh Oyster Mushroom caps (dried mushrooms can be used after hydration)

1 pint whole-fat yogurt

1 cup water

1 cup sliced Groundnuts (see page 69)

salt to taste

Combine 1 pint yogurt with 1 cup water. Heat in a heavy kettle but do not boil. Chop the mushrooms to quarter-size pieces. Add the mushrooms and groundnuts to the warm yogurt and heat to boiling, stirring constantly. Simmer for 10 minutes.

Dehydrated Oyster Mushrooms

Like many mushrooms, Oyster Mushrooms dehydrate very well. Simply dry in an oven at 200°F until the mushrooms are completely dry and snap readily. Immediately after drying, store in a sealed jar in a cool, dark place or freeze. We have stored these for two years with little loss of flavor. Just soak in water to rehydrate and then use as you would fresh.

Collecting

Once you are certain of its identity, cut the caps from the tree. Before collecting, break off a cap and tear it parallel with the gills. If there are tunnels, insect larvae have beaten you to your harvest. Oyster Mushrooms are best just after emerging, when they are firm and free from insect damage, conditions that are found in cool weather. They can even be collected when frozen solid; freezing does not affect their culinary value—just break them off and allow them to defrost.

When transporting your find, take care that bits of dirt and debris from the bark are not mixed with the gills, from which they would be difficult to remove later. Cut off the caps, leaving the tougher, more fibrous stalks. Put the stalks aside; after being thinly sliced they can be slowly cooked with butter and used to flavor soups and casseroles. Sharply rap the cap on a flat surface to remove insects that are commonly found in the gills. (Several insects live in these high-rise arboreal supermarkets and do not necessarily eat your mushrooms.) Like all mushrooms, Oyster Mushrooms are best when they are brushed clean rather than washed. Washing leaches out much of the flavor.

CHICKEN OF THE WOODS

FORTUNATE IS THE FORAGER finding this mushroom! Chicken of the Woods (*Laetiporus sulphureus*), also known as the Sulfur Mushroom, has a distinct color and shape, making it one of the easiest mushrooms to identify. No other shelf fungus with a distinct orange or sulfur color has pores (rather than gills) on its undersurface.

Chicken of the Woods (Laetiporus sulphureus): *This specimen grew so fast that it enveloped the ferns that were growing nearby. This mushroom was growing from buried wood.*

This common fungus can be found just about anywhere on dead or decaying trees. Its appearance is unpredictable, though it may show up in the same place for several years in a row. It is not seasonal but in our experience is usually found in mid- to late summer.

Cut the mushroom from its substrate with a sharp knife and remove the bits of wood and debris that are attached. In very large specimens—up to 10 pounds—the innermost portions may be very woody. These are not toxic but are tough. Unlike many other mushrooms, Chicken of the Woods does not dehydrate well and is best stored frozen after cooking.

RECIPE

Fungus Chicken Fingers

Cut the mushroom into strips 3 inches long and ½ inch wide, trimming tough areas. Sauté in vegetable oil until tender. It is difficult to overcook Chicken of the Woods.

Extra portions can be frozen. When thawed, the strips can be used in soups and casseroles.

CAUTION

THIS IS WIDELY REGARDED as one of the choicest mushrooms for the table. There have, however, been reports of allergies to Chicken of the Woods. Be certain that you are not allergic to this or other mushrooms.

PUFFBALLS

TRUE MUSHROOMS HAVE either gills or pores on the undersurface of their caps; puffballs (species of *Calvatia*) lack both gills and pores. In puffballs, the spores are produced inside and released when the puffball is mature and ruptures. It has been estimated that a single large Giant Puffball has

Giant Puffball
(Calvatia gigantea):
*these Giant Puffballs are
at the ideal stage for eating.*

enough spores to place one on every inch of the state of Pennsylvania. Some people are allergic to the spores. When puffballs are mature they are inedible, though not toxic.

Puffballs are true nitrophiles (that is, organisms that thrive in a nitrogen-rich environment), which explains why they are often most abundant and largest in pastures. The two species that we discuss are both widespread in the United States as well as other countries. These are the Giant Puffball (*Calvatia gigantea*) and the Purple Spore Puffball (*C. cyathiformis*).

Recognizing puffballs is easy. No other fleshy fungus gets as large as these. When you find one, break it open to determine that the interior is white. There is a widespread toxic puffball, but it never exceeds 5 inches in length and has a black, grainy interior. Do not eat any puffball that is gray or dark inside.

The Giant Puffball can be monstrous in size—we have seen specimens 4 feet long and 24 inches in diameter. Giant Puffballs are the mushroom equivalent of Redwood trees—they are among the largest edible mushrooms, easy to identify, not confused with anything toxic, and often abundant. Look for them after rains, when they pop up almost overnight. Both the outside and the interior of the Giant Puffball are white.

Purple Spore Puffballs (C. cyathiformis), *ready to prepare.*

The Purple Spore Puffball is usually smaller and has a leathery, somewhat mottled skin. Unlike the Giant Puffball, this fungus is borne on a cup-shaped structure from which it separates at maturity.

Collecting

Lawns and pastures are good places to find these puffballs, especially in late summer. Those you collect should be firm, in addition to having a white interior. (The inner flesh of the Purple Spore Puffball is sometimes off-white.) Check for insect damage; you'll find it more often in the Purple Spore Puffball. Brush off any dirt.

Both puffballs can be prepared in the same way, although they have different flavors. That of the Purple Spore Puffball is stronger, while the Giant Puffball has a more delicate flavor. Puffballs can be used as a meat substitute in recipes but tend to become mushy when baked.

Puffball Filets

puffballs

cooking oil

salt

Cut the puffballs into medallions not more than ½ inch thick. Place in heated oil and fry *rapidly*—it is very easy to burn puffballs. As soon as the puffball is golden brown, remove it from the oil.

If your puffball is large enough, it is fun to use a cookie cutter. If you want to make, for example, holiday shapes, cut the Puffball into ½-inch slices. Then use cookie cutters to make shapes. These can fried or dehydrated to use later.

Puffballs are especially easy to dehydrate, but they become brittle, so they need to be handled with care

Index of Recipes

Acorn Cakes, Rappahannock, 76
Acorns: *Mok,* 77
Arrowhead Faux Bay Leaf, 69

Basswood leaves: Walnut Wild Plant Chips, 56
Black Locust Fritters, 99
Black Locust Yogurt Dip, 100
Black Walnuts: Walnut Wild Plant Chips, 56
Blueberries: Wild Blueberry Cordial, 121

Cane Crispies, 87
Cattail Corn Dogs, 102
Cattail pollen: Mesopotamian Pollen Paddies, 104
Chicken of the Woods: Fungus Chicken Fingers, 128
Chicory Coffee Substitute, 44
Crunchy Kudzu Leaf Chips, 51
Curly Dock: Salt and Vinegar Strips, 49
Curly Dock Cookies, 47
Curly Dock / Buckwheat Pancakes, 47

Dandelion: Walnut Wild Plant Chips, 56
Day Lily Buds, Pickled, 106
Dehydrated Oyster Mushrooms, 126

Field Garlic Chufa Nuts, 73
Field Garlic Powder, 33
Filé, Groundnut, 71
Filé, Sassafras, 30
Fungus Chicken Fingers, 128

Glasswort: Salt and Vinegar Strips, 49
Groundnut Filé, 71

Indian Strawberry Jam, 113

Kudzu Leaf Chips, Crunchy, 51

Lotus Chips, 66

Manna Grass: Red Sorrel Pilaf, 87
Mesopotamian Pollen Paddies, 104
Mok, 77

Nettle Omelette, 55
Nut Sedge: Field Garlic Chufa Nuts, 73

Oyster Mushrooms, Dehydrated, 126
Oyster Mushroom Soup, 126

Pawpaw Bread, 117
Pickled Day Lily Buds, 106
Pine buds: Walnut Wild Plant Chips, 56
Puffball Filets, 131

Rappahannock Acorn Cakes, 76
Red Sorrel: Walnut Wild Plant Chips, 56
Red Sorrel Pilaf, 89
Red Spruce Aperitif, 39
River Oats Oatmeal, 93
River Oats Pasta, 93

Salt and Vinegar Strips, 49
Sassafras Filé, 30
Sassafras Filé: Black Locust Yogurt Dip, 100
Sauteed Spring Beauty Tubers, 81
Softstem Bulrush, 78
Spring Beauty Tubers, Sauteed, 81
Swamp Bay: Arrowhead Faux Bay Leaf, 69
Swamp Bay Aperitif, 37

Walnut Wild Plant Chips, 56
Wild Blueberry Cordial, 121
Winter Chicory Blanched Leaves, 44